“Wavelets And Wavelet Transforms With Trigonometric Partitions”

José Mauricio Orellana Díaz

How much better to get wisdom than gold! To get understanding is to be chosen rather than silver.Proverbs 16:16 ESV

However, as it is written: “What no eye has seen, what no ear has heard, and what no human mind has conceived” the things GOD has prepared for those who love him. 1Corinthians. 2:9 NIV.

INDEX

0.1 Acknowledgments 5

0.2 Dedications 6

0.3 Summary 7

0.4 Introduction 8

1 Wavelets With Trigonometric Partitions Equations. 9

1.1 Equations Of The Trigonometric Partitions With
Application To Wavelets. 9

1.2 Large Crest Wavelets 28

2 Wavelets With Trigonometric Partitions Of The Circle And Their Transform **35**

2.1 Wavelets Transforms Of Trigonometric Partitions. 35

2.2 Wavelets Transforms Of The Trigonometric
Partitions Of Large Crest. 37

2.3 Wavelets Transforms Of The Trigonometric
Partitions Of The Circular Sector. 43

2.4 Wavelets Transforms Of Trigonometric Partitions Of The Chord. 47

2.5 Wavelets Transforms Of The Trigonometric
Partitions Of The Apothem. 51

2.6 Wavelets Transforms Of Trigonometric Partitions Of The Arrow. 55

2.7 Wavelets Transforms Of Trigonometric Partitions Of Radius Growth. 59
2.8 Application Of Wavelets Transforms Of Trigonometric Partitions To Circular Motion. 63
2.9 Trigonometric Partitions Identities Of Mid-Angles: 70
2.10 Wavelets And Multiple-Wavelet Transforms: 75
2.11 Multiwavelets Transforms As A Function Of Frequency, Angular Velocity, And Time. 83
2.12 Wavelets Of nth Powers: 91
2.13 Wavelets With Damped Simple Harmonic Motion 98
2.14 Annexes: List Of Videos Of The Wavelets With The Theory Of Trigonometric Partition: 121
2.15 Social Networks 122
References 123

0.1 Acknowledgments

I first express my gratitude to God for granting me the wisdom, understanding, and intelligence to develop these mathematical theories, which will be of great benefit to all professionals who seek to analyze physical and mathematical problems through a new methodology.

In His Word we learn that "the fear of the Lord is the beginning of wisdom" (Proverbs 1:7). On many occasions, I did not fully understand why I was developing these theories. In prayer, I asked God to grant me wisdom and, by His grace, I received it. "If any of you lacks wisdom, let him ask of God, who gives to all generously and without reproach, and it will be given to him" (James 1:5). The Proverbs of Solomon teach us: "Get wisdom, get understanding; do not forget my words or turn away from them" (Proverbs 4:5, KJV 1960).

I thank God for His Word, which declares: "Give instruction to a wise man, and he will be still wiser; teach a just man, and he will increase in learning" (Proverbs 9:9). Human wisdom is foolishness before God; He knows the wise and catches them in their craftiness (1 Corinthians 3:19).

It is better to acquire wisdom than fine gold, and to gain understanding is better than silver (Proverbs 16:16). "The heart of the discerning acquires knowledge, and the ear of the wise seeks it out" (Proverbs 18:15). From the depths of my being, I give thanks to God for every blessing, every insight, every measure of wisdom and intelligence He has granted me enabling me to understand many mathematical and physical concepts.

"God is the One who changes times and seasons; He removes kings and establishes kings; He gives wisdom to the wise and knowledge to those who have understanding" (Daniel 2:21, KJV 1960).

0.2 Dedications

As a foretaste of my third mathematical work made possible by the knowledge and wisdom that God has granted me I dedicate this work third and foremost to my Lord and God, who has allowed me to write this third volume. There were moments when I stopped writing these theories and spoke with Him, asking why I was undertaking and developing this work. At times, I was exhausted from so much effort, unable to fully comprehend the depth of what I was doing. I spent countless hours writing in my notebooks, sometimes leaving them unfinished for long periods, yet He continually stirred my heart to continue this work.

Today, having reached this goal for His goodness and favor, I can thank Him infinitely for granting me health and allowing me to write these profound theories on Trigonometric Spirals, Spiral Angles, And Trigonometric Partitions as applied to wavelets.

I also dedicate this work to my children, whom I love with all my heart. When I began this journey of writing these theories. I know that I sacrificed precious time that I could have spent with them, instead investing it in this work. I hope they will read it, learn from it, and witness what they constantly saw me doing tirelessly writing with colored pencils and pens to develop these theories. God has given my precious treasures, wisdom, and they carry the blessing of becoming wise, intelligent, and discerning in this world, making good use of their time each day.

The glory of God is given for what He has done in me. I thank Him for His love and infinite mercy over my life, for it is He who has helped me reach the final goal of completing this book presenting these new and fascinating theories in the field of mathematical physics.

0.3 Summary

These mathematical equations represent a new methodology that complements existing wavelet theories. They expand the family of wavelets by generating different types through the equations of the radius as a function of the trigonometric partitions of a circle's components such as the apothem, arrow, chord, circular sector, radial growth, or their equivalent expressions, where the radius is defined as the sum of the arrow and the apothem, as well as any other equations that may be defined in relation to the radius of a circle or a regular polygon.

Using the components of each element of the circle, wavelets can be generated without relying on existing theories. Under this new methodology, the current foundational theories such as complex numbers, Fourier series, Fourier and Laplace transforms, or the SINC trigonometric function are not required. This methodology provides a fundamental understanding of how a wavelet is born, especially the initial radius wavelet, expressed as a function of the secant or cosecant for large crest wavelets. Notably, it is not necessary to know its final radius, since the equation of the initial radius is defined in terms of trigonometric partitions and the angle, or its equivalent transformed angle equations.

With this methodology, all standard trigonometric identities can be redefined as trigonometric identities expressed in terms of the angle and trigonometric partitions, allowing the final equations to be simplified for solving both classical and quantum mathematical and physical problems.

There are two approaches to defining equations for generating wavelets. The first defines the radius as inversely proportional to the sine or cosine of its angle (that is, its equivalent secant or cosecant) without using trigonometric partitions. The second employs the spiral radius equations in relation to trigonometric partitions and their spiral angles, as demonstrated in the equations for large crest wavelets. Finally, one learns how to generate wavelets through the transformations of equations originally written in terms of the angle, using their mathematically equivalent angle based equations.

0.4 Introduction

The Theory of Spiral Angles, Spirals, and Trigonometric Partitions is an extension to current theories of spirals and wavelets. It offers applications to spirals that do not have a defined mathematical formula, such as the Spiral of Theodorus, as well as to spirals with established formulas, including the Archimedean, Logarithmic, and Hyperbolic spirals.

The spiral angle is a special type of angle like a fingerprint whose calculations inherently contain all the properties of a circle or a regular polygon.

Using this theory, triangles can be constructed and analyzed using all fundamental trigonometric identities. Furthermore, this framework has made it possible to derive a general trigonometric identity for continuous mean partitions of a circle. Through these mean trigonometric identities, one can compute the integrals of such partitions, generate their graphs, and analyze wavelets derived from them.

Trigonometric spirals follow their own governing laws. When analyzed at infinity, these spirals yield simplified equations, allowing us to determine whether or not they produce spirals. This also makes it possible to measure deviations between standard trigonometric spirals and their circular shape behavior of the trigonometric spiral.

One key application of this theory is in generating models for economic systems with spiral growth. Finally, one of the most significant contributions of this work is Theory of Trigonometric Partitions and its fundamental equations,which can be applied as an entirely new methodology for the construction of wavelets, the study of the Riemann Hypothesis, Goldbach's Conjecture, the distribution of Prime Numbers, and other mathematical phenomena such as harmonics, String theory, Quantum physics, music, sound, and signal analysis,etc.

Chapter 1

Wavelets With Trigonometric Partitions Equations.

1.1 Equations Of The Trigonometric Partitions With Application To Wavelets.

In 1984, Haar developed a set of methods and theories, which would eventually lead to the birth of the notion of ondelette; in English, they are translated as "wavelet." (Fernández, 2019)

All the standard basic trigonometric identities can be expressed as functions of the equations of spiral angles, spirals, and trigonometric partitions.

Figure 1.1: Diagram of Trigonometric Partitions of the Circle.

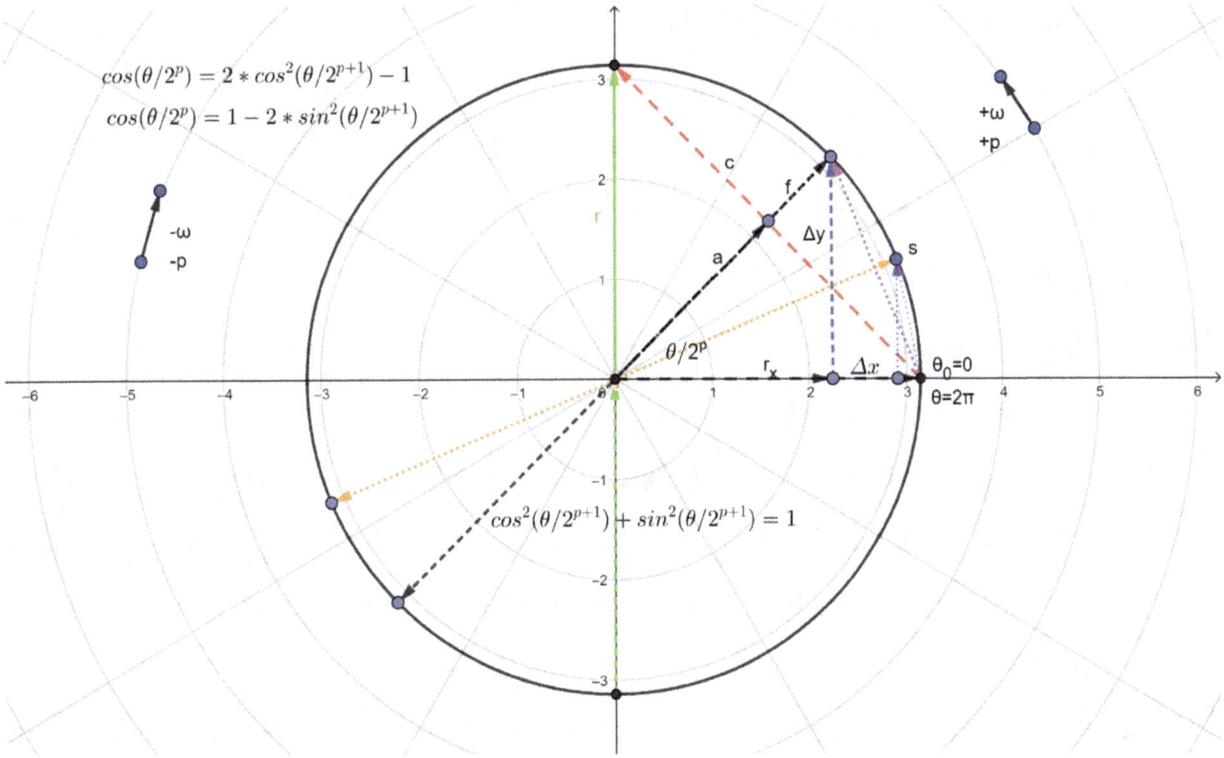

Note:With this scheme, you can obtain the trigonometric identities of mean angles in general and the components of the circle based on the Trigonometric Partitions.

With the transformation of the basic equations of the components of a circle, based on the trigonometric partitions, such as; with $p \geq 1$:

The circular sector: $s = r * \theta$. For small angles $\theta \approx 2 * \sin(\theta/2)$, with $\theta = 2^p * \sin(\theta/2^p)$.

The chord: $c = 2 * r * \sin(\theta/2)$.

The apothem: $a = r * \cos(\theta/2)$.

The arrow: $f = 2 * r * \sin^2(\theta/4)$.

The growth of the radius: $\Delta r = 2 * r * \sin^2(\theta/2)$.

The most valuable equations are obtained from the trigonometric partitions of the components of a circle. With which we can generate a great variety of waves and wavelets. Expanding the family of existing wavelets. Likewise, we can do the same; create wavelets with the derivatives of the equations of the trigonometric partitions of the radius for each of the components of the circle.

The Theory of Spiral Angles, Spirals, And Trigonometric Partitions teaches us a new methodology to make wavelets without using complex numbers, Fourier series, Fourier and Laplace transforms, or the SINC trigonometric function. Especially the polar **Y** coordinate wavelet, where complex numbers are required, or the Fourier series and its transforms.

Additionally, for each element that is part of a circle; such as the circular sector, the chord, the apothem, the arrow, and the growth of the radius, it was discovered that for each equation its first derivative of the radius as a function of the angle will always be equal to the derivative of its component of the circle, so consequent; $dr/d(\theta) = s, dr/d(\theta) = c,$ $dr/d(\theta) = a, dr/d(\theta) = f, dr/d(\theta) = \Delta x$.(Orellana Díaz, 2023)

With this new methodology for generating wavelets, we can study the effect produced by each of the variables in the equations, either individually or in combination, by observing the resulting behavior in the plots depending on the type of equation being used.

In each of the following equations, we can use the trigonometric partitions corresponding to each of the elements of the circle.

If we substitute the equations of the trigonometric partitions of the radius of each element of the circle, into the following equations; is equivalent to obtaining the derivative of the angle as a function of the radius of each of the elements of the circle.

The following equations: are the equations of the elements of a circle based on the trigonometric partitions **p**, for example **s**: is the circular sector, **c**: is the chord,**a**: is the apothem, **f**: is the arrow, Δr or Δx: is the growth of the radius or x.

1. $s = 2^p * r * \sin(\theta/2^p)$ where p is -$\infty \leq p \geq +\infty$ in the infinite circular sector tends to the multiple of the radius times the angle θ; $s = r * \theta$.
2. $c = 2 * r * \sin(\theta/2^p)$ where p is -$\infty \leq p \geq +\infty$ at infinity the chord tends to zero.
3. $a = r * cos(\theta/2^p)$ where p is -$\infty \leq p \geq +\infty$ at infinity the apothem tends to the radio.
4. $f = 2 * r * \sin^2(\theta/2^{p+1})$ where p is -$\infty \leq p \geq +\infty$ in infinity the arrow tends to zero.
5. $\Delta x = \Delta r = 2 * r * \sin^2(\theta/2^p)$ where p is -$\infty \leq p \geq +\infty$ in infinity the increase of the radius tends to zero.

If we apply the integrating factor of the circular sector **s**; the radius will be obtained **r** as a function of the trigonometric partitions of the circular sector; so that it has a trigonometric spiral growth and we analyze based on the unit circular sector, we will have the following equation of its integration factor:

$r = r_0 * e^{\int_0^\theta (s(\theta)/r)\, d\theta}$,

$s(\theta) = 2^p * r * \sin(\theta/2^p)$ therefore;

$\int_0^\theta (s(\theta)/r) * d\theta = \int_0^\theta 2^p * \sin(\theta/2^p)\, d\theta = 2^{2*p+1} * \sin^2(\theta/2^{p+1})$ substituting will obtain the integral defined with $\theta_0 = 0$, $r = r_0 * e^{2^{2*p+1} * \sin^2(\theta/2^{p+1})}$. In this equation when the growth of the radius is constant Example:

$r = r_0 * e^{2^{2*p+1} * \sin^2(\theta/2^{p+1})}$; are the trigonometric partitions that produce the generation of wave function and wavelets.

Substituting $r = r_0 * e^{2^{2*p+1} * \sin^2(\theta/2^{p+1})}$ into:

$s = 2^p * r * \sin(\theta/2^p)$ where p is -$\infty \leq p \geq +\infty$ at infinity the circular sector tends to the multiple of the radius times the angle θ; $s = r * \theta$.

The radius equation that relates the circular sector to the trigonometric partitions will be obtained in: $s = 2^p * r_0 * \sin(\theta/2^p) * e^{2^{2*p+1}*\sin^2(\theta/2^{p+1})}$.

Therefore, with the final radius r its amplitude can be obtained. Observe what happens in the graph made in GeoGebra, when the two sliders in the GeoGebra program begin to animate. The waves begin to multiply on the graphs. In these equations can be used depending on angular velocity, time, frequency, etc.

$\theta = \omega * t, \theta = 2 * \pi * f * t, \theta = k * x - \omega * t, k = 2 * \pi/\lambda$, etc.

All these equations are differentiable and integrable. Whether they are normal or partial derivatives depends on the case where they are being applied, with which wavelets can be generated.

Solving for the initial radius, the function of the initial radius about the angle and its polar coordinates; $r_0 = r * e^{-2^{2*p+1}*\sin^2(\theta/2^{p+1})}$,

$r_{0x} = r * \cos(k * \theta) * e^{-2^{2*p+1}*\sin^2(\theta/2^{p+1})}$,

$r_{0y} = r * \sin(k * \theta) e^{-2^{2*p+1}*\sin^2(\theta/2^{p+1})}$.

The variable **k** arises from the spiral based analysis and is the one that generates the waveforms of the wavelets; this is why it is used exclusively in the component variables.The use of the constant **k** in trigonometric spirals should not be confused with the one that relates to the wavelength $k = 2 * \pi/\lambda$.

But sometimes this is used to show that by making manual changes to the trigonometric partitions, wavelets continue to be generated.But this does not always happen with some equations.

Initially, I developed the wavelets on an Excel sheet, where I was only able to construct the right half but could not obtain the left half. Upon analyzing the situation, I discovered that a factor **(k)** was required to develop the angular values in series and compute their summations, allowing the generation of the left half of the wavelet. This left half represents the imaginary part of the wave, which is similar to the wavelet corresponding to the **(y)**-coordinate most can only develop this using complex numbers. In contrast, the wavelet of the **(x)**-coordinate represents the real part.

Figure 1.2: Equation of Trigonometric Partitions of the Circular Sector.

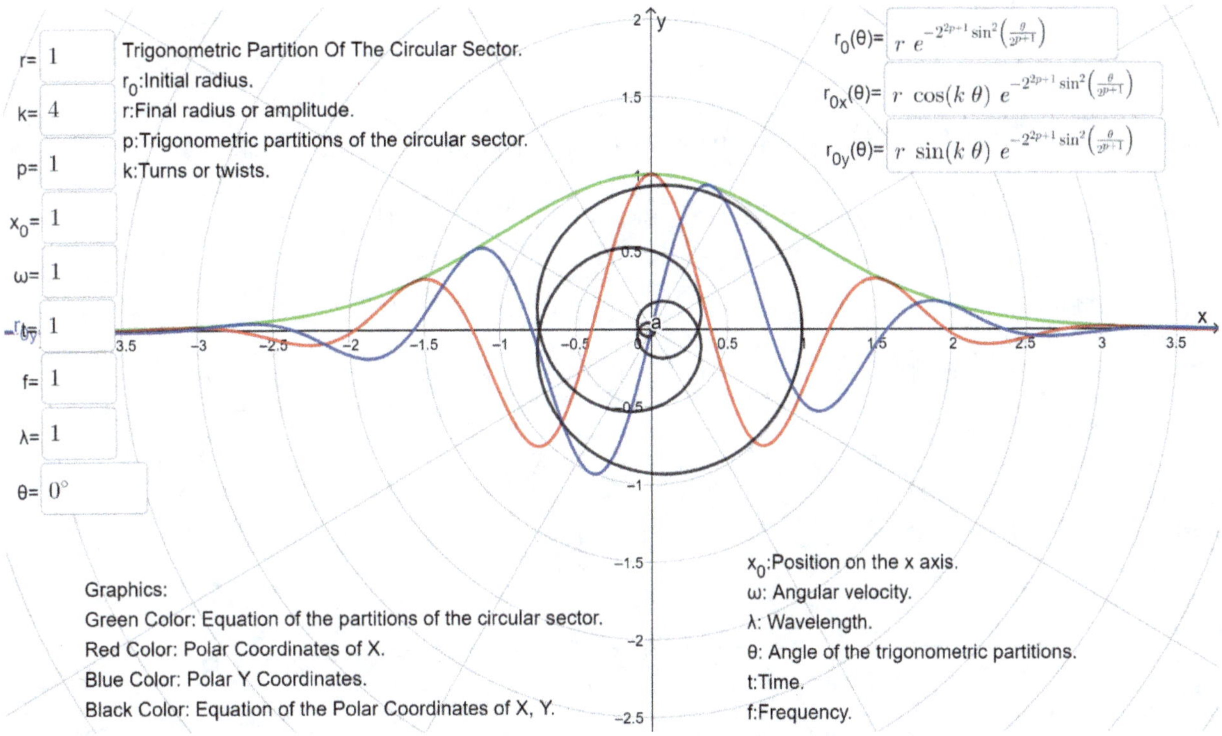

Note: This graph shows how the wavelets are generated with the equation of the initial radius as a function of the angle.

In these equations **n** is the number of circles and **p** the partitions; they can take any value. r_0; is the wave equation for real numbers of decrease (Initial Radius) or growth of the radius (Final Radius), since Its exponential function can be positive or negative.

Note: Example with k = 4 and n = 4. (**k**:number of turns,**n**:number of circles or levels). With the initial radius negative ($-r$) the graphs are inverted.

$r_{0x} = r * e^{-2^{2*p+1}*n*\sin^2(\theta/2^{p+1})} * \cos(k*\theta) = r_0 * \cos(k*\theta).$

$r_{0y} = r * e^{-2^{2*p+1}*n*\sin^2(\theta/2^{p+1}} * \sin(k*\theta) = r_0 * \sin(k*\theta).$

Figure 1.3: Equation of Trigonometric Partitions of the Circular Sector.

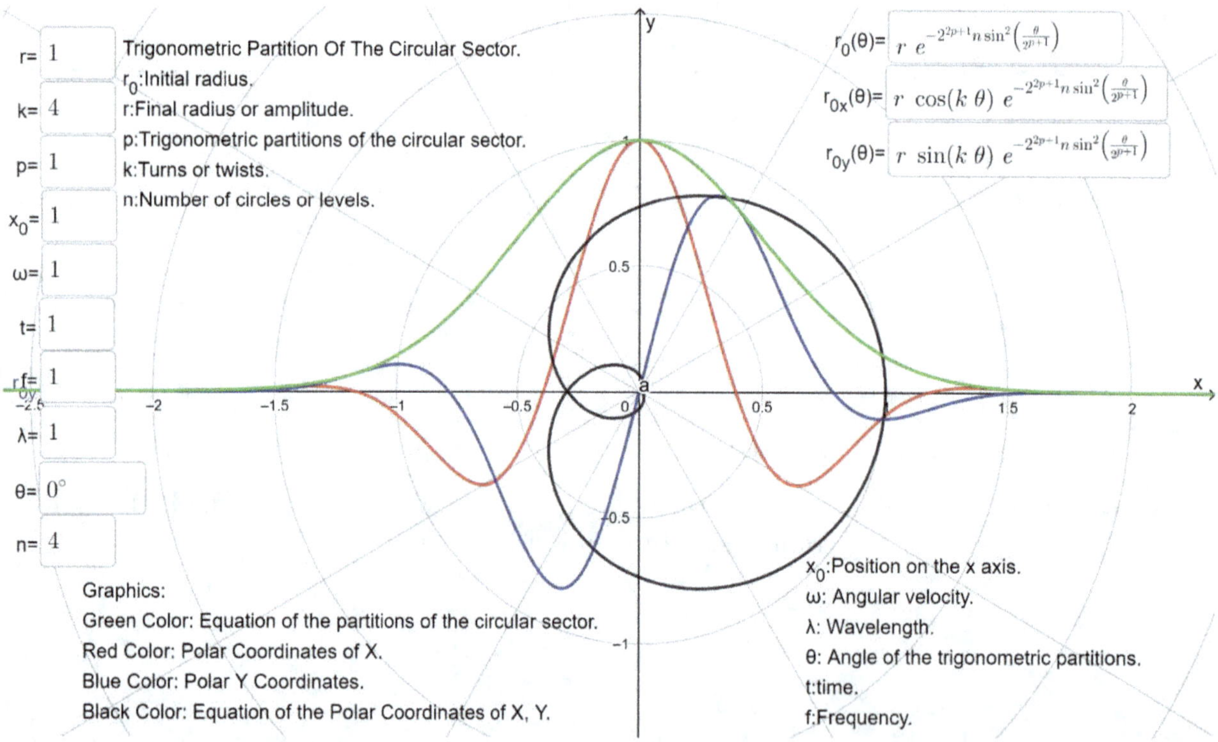

Note:This graph shows how the wavelets are generated with the equation of the initial radius as a function of the angle with n levels or circles.

Other types of wavelets can be generated, if we use the powers of these same equations, such as those generated in the patterns of interferences. $r_0^2 = r^2 * e^{-2^{2*p+2}*\sin^2(\theta/2^{p+1})}$,

$r_{0x}^2 = r^2 * \cos^2(k*\theta) * e^{-2^{2*p+2}*\sin^2(\theta/2^{p+1})} = r_0^2 * \cos^2(k*\theta)$.

$r_{0y}^2 = r^2 * \sin^2(k*\theta) * e^{-2^{2*p+2}*\sin^2(\theta/2^{p+1})} = r_0^2 * \sin^2(k*\theta)$.

Figure 1.4: Equation of Trigonometric Partitions of the Circular Sector.

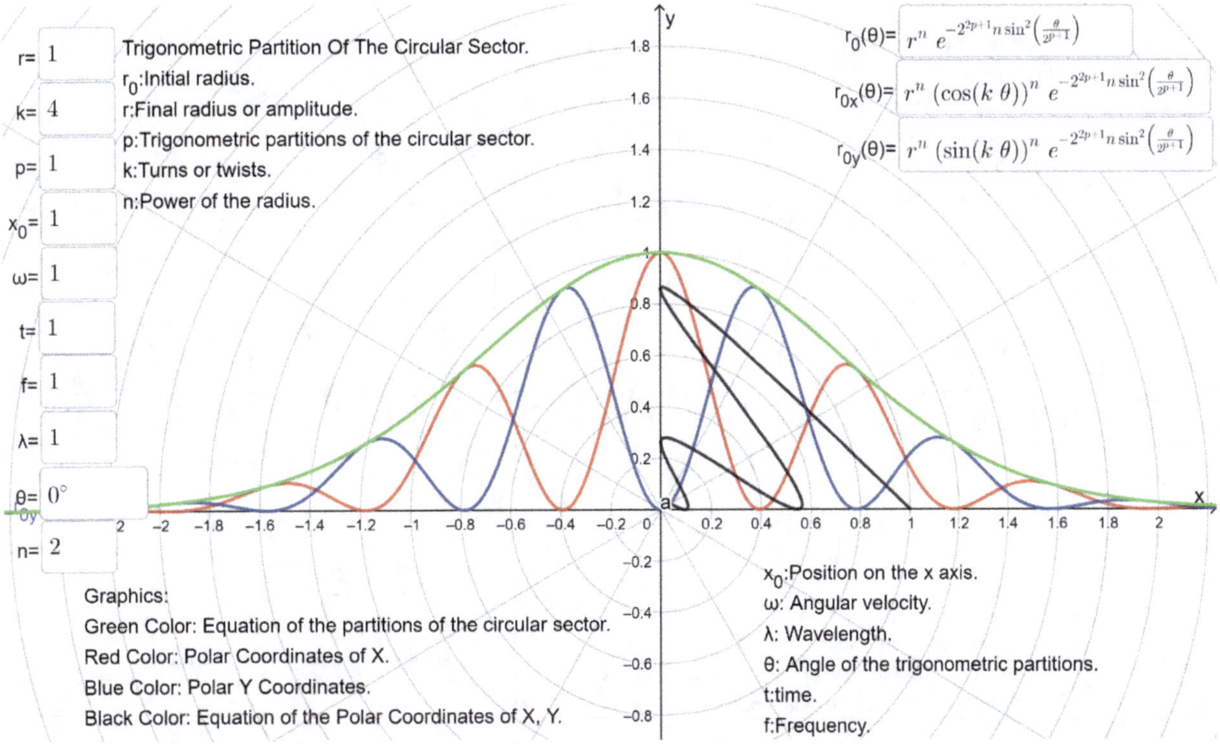

Note:This graph shows how the wavelets are generated with the equation of the initial radius as a function of the angle with the nth power of the radius.

With these equations, you can use revolutions per minute RPM, frequency, angular velocity, where;$\theta = \omega * t, \theta = 2 * \pi * f * t$, or $\theta = k * x - \omega * t$, with $k = 2 * \pi / \lambda$ etc. Since their origin is in a polar circular plane.

Any arithmetic operation that we apply to these equations wavelets will always be generated.

Figure 1.5: Equation of Trigonometric Partitions of the Circular Sector.

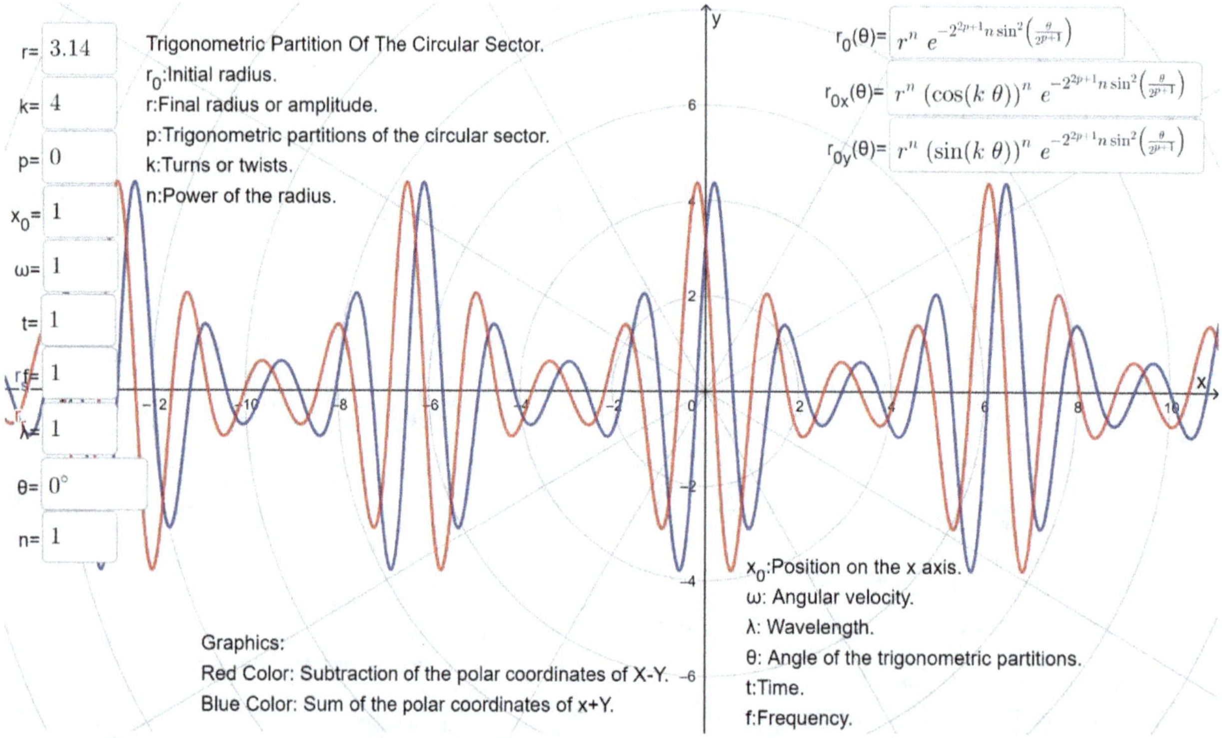

Note:This graph shows how the wavelets are generated with the sum equations and subtraction of the x and y polar coordinates.

Figure 1.6: Equation of Trigonometric Partitions of the Circular Sector.

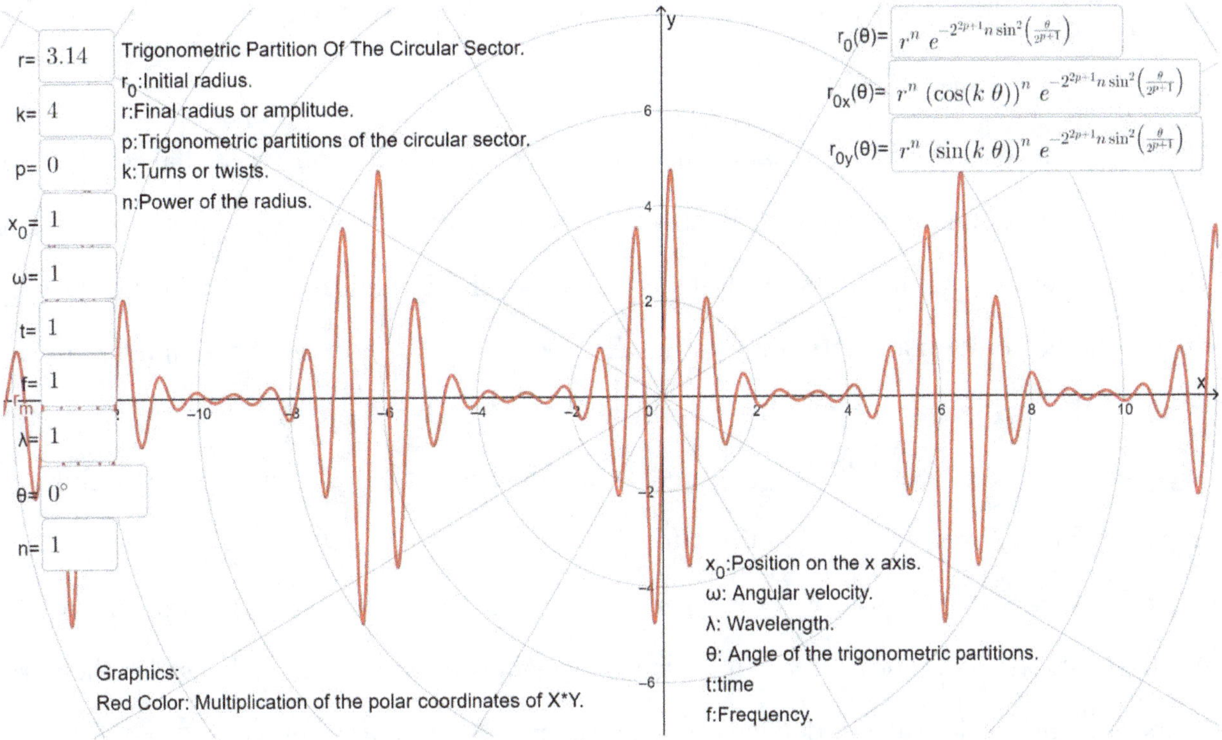

Note:This graph shows how the wavelets are generated with the equations for multiplying the polar coordinates x and y.

Note: For more information, you can watch videos on my YouTube channel.

Applying, the integrating factor; to the chord to obtain the final radius r; so that it has a trigonometric spiral growth and we analyze based on the unit chord, we will have the following equation of its integration: $r = r_0 * e^{\int_0^{\theta}(c(\theta)/r)\,d\theta}$,

$c(\theta) = 2 * r * \sin(\theta/2^p)$ therefore,

$\int_0^{\theta}(c(\theta)/r) * d\theta = \int_0^{\theta} 2 * \sin(\theta/2^p)\,d\theta = 2^{p+2} * \sin^2(\theta/2^{p+1})$,

substituting you will get the integral defined with $\theta_0 = 0$: $r = r_0 * e^{2^{p+2} * \sin^2(\theta/2^{p+1})}$.

Substituting $r = r_0 * e^{2^{p+2} * \sin^2(\theta/2^{p+1})}$ into:

$c = 2 * r * \sin(\theta/2^p)$ where p is -$\infty \leq p \geq +\infty$ at infinity the chord tends to zero.

The radius equation that relates the chord to the trigonometric partitions will be obtained in: $c = 2 * r_0 * \sin(\theta/2^p) * e^{2^{p+2} * \sin^2(\theta/2^{p+1})}$.

Solving for the initial radius, the function of the initial radius in relation to the angle and its polar coordinate : $r_0 = r * e^{-2^{p+2} * \sin^2(\theta/2^{p+1})}$,

$r_{0x} = r * \cos(k * \theta) e^{-2^{p+2} * \sin^2(\theta/2^{p+1})}$,

$r_{0y} = r * \sin(k * \theta) e^{-2^{p+2} * \sin^2(\theta/2^{p+1})}$.

This equation is very similar to that of the circular sector, here it is observed that when both equations tend to infinitely the chord and the circular sector they are equal. In this equation when the final radius is left constant, the partitions are that generate exponential sine waves. And with the radio final r, its amplitude can be obtained. Observe what happens in the graph made in GeoGebra, when the two sliders in the GeoGebra program begin to animate. The waves begin to multiply on the graphs. In these equations if can use, depending on angular velocity, time, frequency, etc.

$\theta = \omega * t, \theta = 2 * \pi * f * t$, or $\theta = k * x - \omega * t$, with $k = 2 * \pi/\lambda$. All these equations are differentiable and integrable. With which wavelets can be generated. Whether they are normal or partial derivatives depends on the case where they are applying.(José Mauricio Orellana Díaz, n.d.-c)

Figure 1.7: Equation Of Trigonometric Partitions Of The Chord.

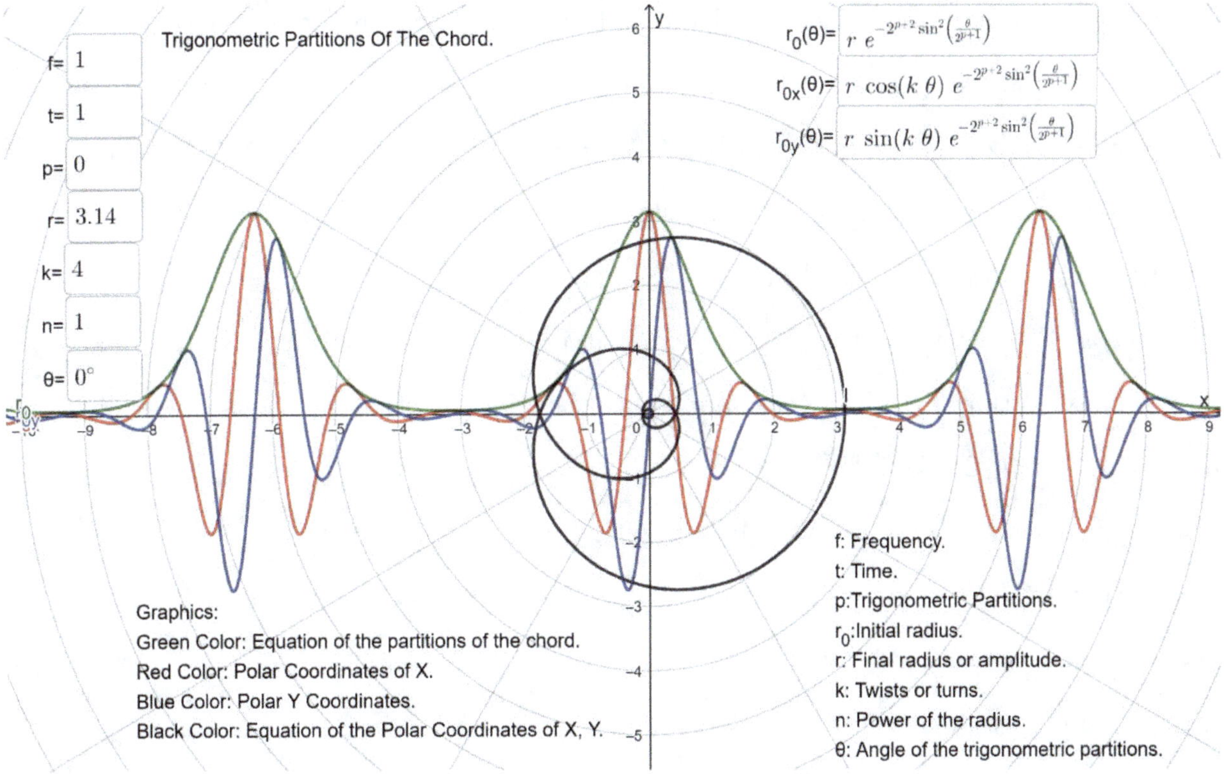

Note:This graph shows how the wavelets are generated with the equations of the initial radius as a function of the angle.

Applying the integrating factor to the apothem to obtain the final radius r; for that has a trigonometric spiral growth and we analyze with based on the unitary apothem, we will have the following equation of its integration: $r = r_0 * e^{\int_0^\theta (a(\theta)/r)\,d\theta}$,

$a(\theta) = r * \cos(\theta/2^p)$ Therefore, $\int_0^\theta (a(\theta)/r) * d\theta = \int_0^\theta \cos(\theta/2^p)\,d\theta = 2^p * \sin(\theta/2^p)$ substituting; the definite integral will be obtained with $\theta_0 = 0$: $r = r_0 * e^{2^P * \sin(\theta/2^p)}$.

Substituting $r = r_0 * e^{2^P * \sin(\theta/2^p)}$ into:

$a = r * cos(\theta/2^p)$ where p is -$\infty \leq p \geq +\infty$ at infinity, the apothem tends to the radio.

The radius equation that relates the apothem to the trigonometric partitions will be obtained in: $a = r_0 * cos(\theta/2^p) * e^{2^P * \sin(\theta/2^p)}$.

These equations are very similar. Observe with $k = 3, p = 2, r = 0.1$,

$r_0 = r * e^{-2^P * \sin(\theta/2^p)}$.

$r_{0x} = r * \cos\,(k * \theta) * e^{-2^P * \sin(\theta/2^p)}$,

$r_{0y} = r * \sin\,(k * \theta) * e^{-2^P * \sin(\theta/2^p)}$

In this equation when the final radius is left constant, They are the partitions that generate sine waves. And with the final radius r, its amplitude can be obtained. Observing what happens in the graph made in GeoGebra, when the two sliders of the GeoGebra program they start to animate. The waves begin to multiply on the graphs. These equations can be used as a function of the angular velocity, time, and frequency, etc.

$\theta = \omega * t, \theta = 2 * \pi * f * t$, or $\theta = k * x - \omega * t$, with $k = 2 * \pi/\lambda$. All these equations are differentiable and integrable. With which wavelets can be generated.

Figure 1.8: Equation Of The Trigonometric Partitions Of The Apothem.

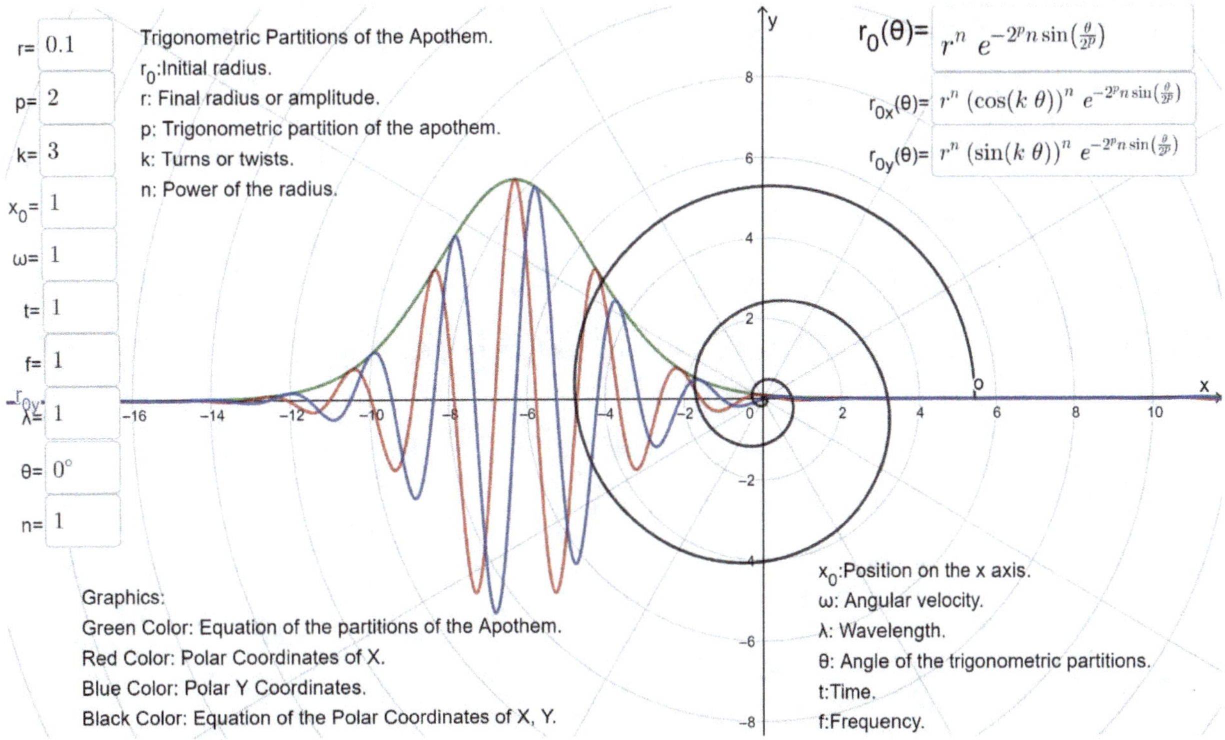

Note:This graph shows how the wavelets are generated with the equations of the initial radius as a function of the angle.

Applying the integrating factor to the arrow to obtain the final radius r; so that it has a trigonometric spiral growth we analyze based on the unit arrow, we will have the following equation of its integration: $r = r_0 * e^{\int_0^\theta (f(\theta)/r)\, d\theta}$,

$f(\theta) = 2*r*\sin^2(\theta/2^{p+1})$ therefore, $\int_0^\theta (f(\theta)/r) * d\theta = \int_0^\theta 2*\sin^2(\theta/2^{p+1})\, d\theta = \theta - 2^p * \sin(\theta/2^p)$ substituting will obtain the integral defined with $\theta_0 = 0$. Where θ is in radians.

$r = r_0 * e^{\theta - 2^P * \sin(\theta/2^P)}$.

Substituting $r = r_0 * e^{\pi*\theta/180 - 2^P * \sin(\theta/2^P)}$ into:

$f = 2*r*\sin^2(\theta/2^{p+1})$ where p is $-\infty \leq p \geq +\infty$ in infinity, the arrow tends to zero.

The radius equation that relates the arrow to the trigonometric partitions will be obtained in: $f = 2*r_0*\sin^2(\theta/2^{p+1}) * e^{\pi*\theta/180 - 2^P * \sin(\theta/2^P)}$.

Observe with $k = 2, p = 2, r = 0.1$, $r_0 = r * e^{2^P * \sin(\theta/2^P) - \theta}$.

$r_{0x} = r * \cos(k*\theta) * e^{2^P * \sin(\theta/2^P) - \theta}$. and $r_{0y} = r * \sin(k*\theta) * e^{2^P * \sin(\theta/2^P) - \theta}$.

Where θ is in degrees. In this equation when the final radius is left constant, the partitions are that generate exponential waves. And with the final radius r, you can get its amplitude. Observe what happens in the graph made in GeoGebra, when the two sliders of the GeoGebra program start to animate. The waves begin to multiply on the graphs. These equations can be used as a function of the angular velocity, the time, and frequency, etc.

$\theta = \omega * \mathrm{t}, \theta = 2 * \pi * \mathrm{f} * \mathrm{t}$, or $\theta = \mathrm{k} * \mathrm{x} - \omega * \mathrm{t}$, with $\mathrm{k} = 2 * \pi / \lambda$. All these equations are differentiable and integrable. With which wavelets can be generated.

Figure 1.9: Equation Of Trigonometric Partitions Of The Arrow.

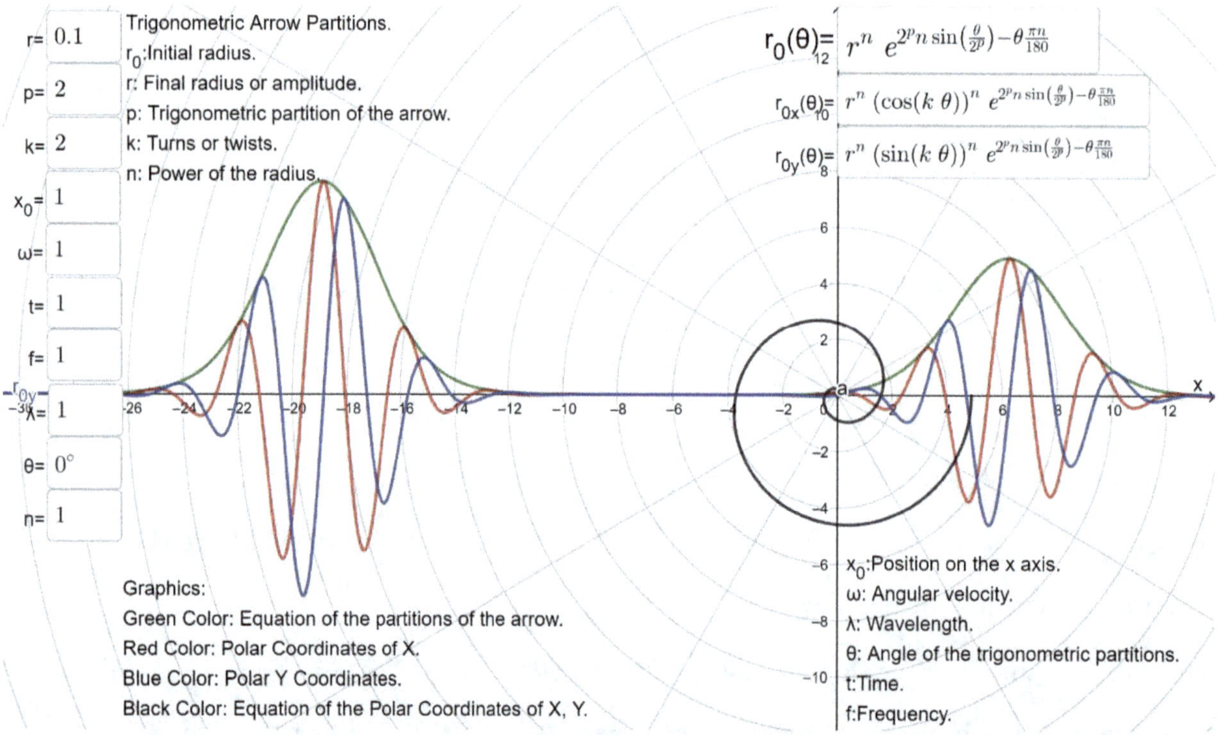

Note: This graph shows how wavelets of different sizes are generated with the equations of the initial radius as a function of the angle. The polar coordinates are a spiral.

Applying the integrating factor to the growth of the radius $\Delta r = \Delta x$ to obtain the final radius r; so that it has a trigonometric spiral growth and it is analyzed based on the delta of r, or unitary x. The following equation will be obtained for its integration: $\mathrm{r} = \mathrm{r}_0 * \mathrm{e}^{\int_0^\theta (\Delta \mathrm{x}(\theta)/\mathrm{r})\, \mathrm{d}\theta}$.

$\Delta \mathrm{x}(\theta) = \Delta \mathrm{r}(\theta) = 2 * \mathrm{r} * \sin^2(\theta/2^{\mathrm{p}})$ therefore,

$\int_0^\theta (\Delta x(\theta)/r) * d\theta = \int_0^\theta 2 * \sin^2(\theta/2^p)\, d\theta = \theta - 2^{p-1} * \sin(\theta/2^{p-1})$.

Substituting will obtain the integral defined with $\theta_0 = 0$. Where θ is in radians.

$r = r_0 * e^{\theta - 2^{p-1} * \sin(\theta/2^{p-1})}$.

Substituting $r = r_0 * e^{\pi * \theta/180 - 2^{p-1} * \sin(\theta/2^{p-1})}$ into:

$\Delta x = \Delta r = 2 * r * \sin^2(\theta/2^p)$ where p is $-\infty \leq p \geq +\infty$ in infinity, the increase of the radius tends to zero.

The radius equation that relates the growth of the radius to the trigonometric partitions will be obtained in: $\Delta x = \Delta r = 2 * r_0 * \sin^2(\theta/2^p) * e^{\pi * \theta/180 - 2^{p-1} * \sin(\theta/2^{p-1})}$. Where θ is in degrees

Observe with $k = 2, p = 3, r = 0.1$, $r_0 = r * e^{2^{p-1} * \sin(\theta/2^{p-1}) - \theta}$.

$r_{0x} = r * \cos(k * \theta) * e^{2^{p-1} * \sin(\theta/2^{p-1}) - \theta}$.

$r_{0y} = r * \sin(k * \theta) * e^{2^{p-1} * \sin(\theta/2^{p-1}) - \theta}$.

In this equation, when the final radius is constant; are the partitions that generate exponential waves. And with the final radius r, we can obtain its amplitude. Observe what happens in the graph made in GeoGebra, when the two sliders in the GeoGebra program begin to animate. The waves begin to multiply on the graphs. In these equations whether they can be used as a function of angular velocity, time, frequency, etc.

$\theta = \omega * t, \theta = 2 * \pi * f * t$, or $\theta = k * x - \omega * t$, with $k = 2 * \pi/\lambda$.

All these equations are differentiable and integrable.With which wavelets can be generated.

Figure 1.10: Equation of the Trigonometric Partitions Of The Growth Of The Radius.

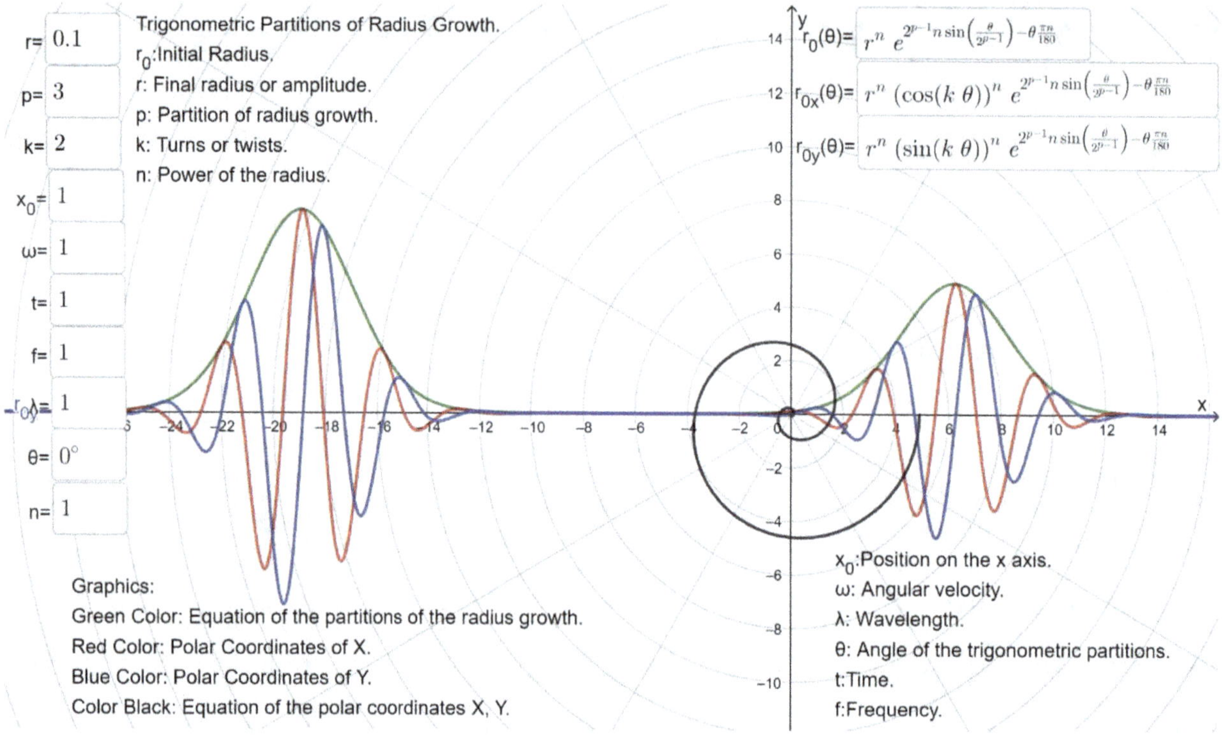

Note:This graph shows how wavelets of different sizes are generated with the equations of the initial radius as a function of the angle. The polar coordinates are a spiral.

Therefore, these equations can be written in a general form: In order to move it in the Cartesian plane at different points. for the decrease of the radius (-), the wavelets are mostly small, or for the growth (+) of the same, the wavelets are mostly large.

$r_0 = r * e^{-2^{2*p+1} * \sin^2(\theta/2^{p+1})} + g$. This is the wave equation for numbers real.

$r_{0x} = r * e^{-2^{2*p+1} * \sin^2(\theta/2^{p+1})} * \cos(k * \theta - \varphi) + g * \cos(k * \theta - \varphi)$. With decreasing radii.

$r_{0y} = r * e^{-2^{2*p+1} * \sin^2(\theta/2^{p+1})} * \sin(k * \theta - \varphi) + g * sin(k * \theta - \varphi)$. With decreasing radii.

The Fundamental Equations of Trigonometric Partitions:

If we integrate these equations for their angles, we obtain equations similar, with $\theta_0 = 0$:

Considering that the decrease of the radius is in a trigonometric spiral form. As a result of the radius of a circle, the fundamental equation can be substituted in the equations:

All these equations are differentiable and integrable; Whether derived normal or partial depending on the case where they are being applied.

$r_0 = r * e^{-2^{2*p+1} * \sin^2(\theta/2^{p+1})}$. Where the interval of the partitions p is $-\infty \leq p \geq +\infty$

1.$s_0 = 2^p * r_0 * \sin(\theta/2^p)...s_0 = 2^p * r * \sin(\theta/2^p) * e^{-2^{2*p+1} * \sin^2(\theta/2^{p+1})}$, $s_0 = s * e^{-2^{2*p+1} * \sin^2(\theta/2^{p+1})}$.

2.$c_0 = 2 * r_0 * \sin(\theta/2^p)...c_0 = 2 * r * \sin(\theta/2^p) * e^{-2^{2*p+1} * \sin^2(\theta/2p+1}$, $c_0 = c * e^{-2^{2*p+1} * \sin^2(\theta/2^{p+1})}$.

3.$a_0 = r_0 * \cos(\theta/2^p)...a_0 = r * \cos(\theta/2^p) * e^{-2^{2*p+1} * \sin^2(\theta/2^{p+1})}$, $a_0 = a * e^{-2^{2*p+1} * \sin^2(\theta/2^{p+1})}$.

4.$f_0 = 2 * r_0 * \sin^2(\theta/2^{p+1})...f_0 = 2 * r * \sin^2(\theta/2^{p+1}) * e^{-2^{2*p+1} * \sin^2(\theta/2^{p+1})}$, $f_0 = f * e^{-2^{2*p+1} * \sin^2(\theta/2^{p+1})}$.

5.$\Delta x_0 = \Delta r_0 = 2 * r_0 * \sin^2(\theta/2^p)...\Delta r_0 = 2 * r * \sin^2(\theta/2^p) * e^{-2^{2*p+1} * \sin^2(\theta/2^{p+1})}$,

$\Delta r_0 = \Delta r * e^{-2^{2*p+1} * \sin^2(\theta/2^{p+1})}$.Other equations can be derived with $r = a + f$,

$r = r_x + \Delta r$, with r_x They can be related to three equations of trigonometric partitions: $r = r_x + r_{sx}$ of the x component of the circular sector, $r = r_x + r_{cx}$ of the x component of the chord, and the last one related to the diameter for its diametral cuts of the circle with $d = 2 * r$.

Considering that the initial radius decreases in a spiral with r constant and a constant angle θ, with the number of circles **(n)** we can calculate its initial radius.

$r_0 = r * e^{-2^{2*p+1} * n * \sin^2(\theta/2^{p+1})}$; With this equation, we can calculate the frequency of the earth. And apply it to the double-slit experiment. Example of the number of circles **n** = 1; corresponds to the first circle and so on. Therefore;**n** corresponds to the number of concentric circles n=1,2,3...

Figure 1.11: Equation Of Trigonometric Partitions Of The Circular Sector By Levels.

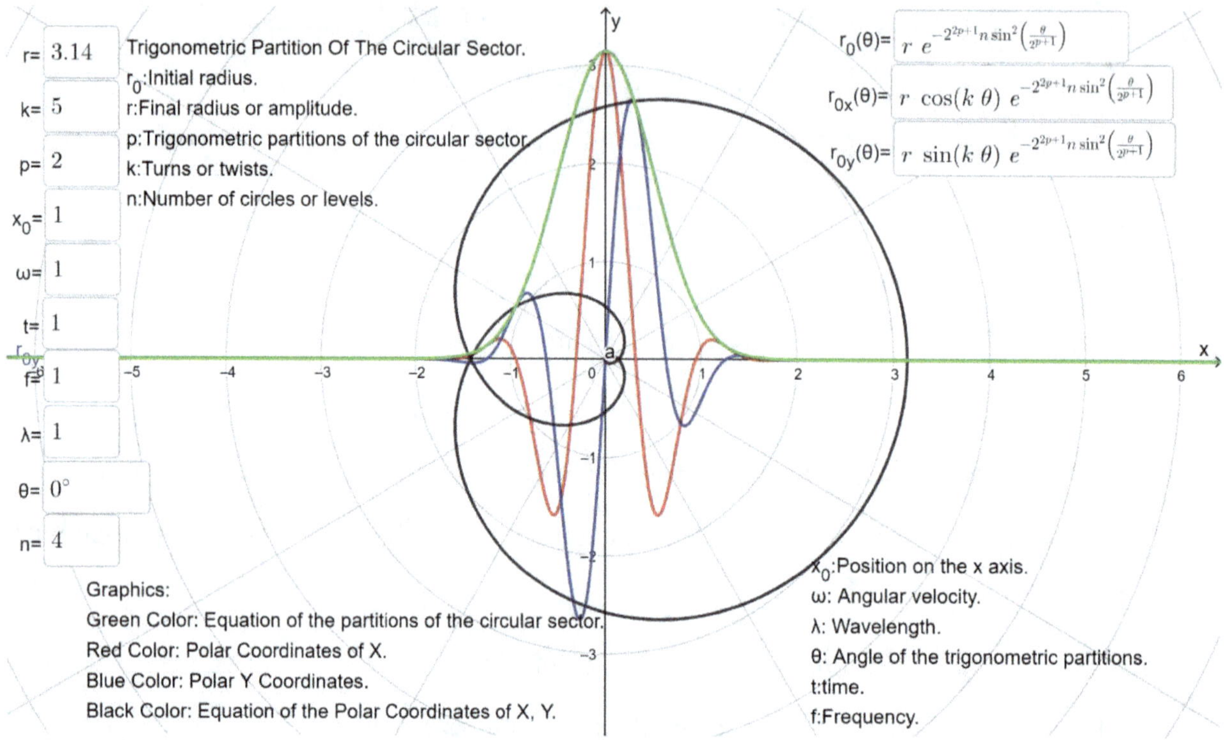

Note:This graph shows how the wavelets are generated by the n levels with the equations of the initial radius as a function of the angle.

1.2 Large Crest Wavelets

Large central crest wavelets are special wavelets which are independent of the final radius; they are generated by the cosecant or secant function, depending on which trigonometric partitions equation is being used. To generate this type of wavelets, we can use all the trigonometric partitions equations; likewise, all the basic trigonometric identities can be converted into trigonometric partition functions, which are useful in case derivations of the trigonometric partitions equations are performed. To generate these wavelets, we can use the radius in spiral form without trigonometric partitions, or likewise, use the radius in spiral form with trigonometric partitions in this way we will generate two equations of the initial radius independent of the final radius and dependent on the trigonometric partitions and the angle or its transformations for the angle equation. The variables **k** and **p** are the multiplying factors that make the crest of the wavelet large. With the trigonometric partitions equations, we can create wavelets by adding, subtracting, multiplying, dividing two trigonometric partitions equations, raising them to a power, differentiating, integrating, and introducing into a mathematical equation the trigonometric partitions equations. It is even possible to manually change the variables of a trigonometric partitions equation that generates wavelets its results will be the generation of another wavelet.

If we use the equation of the radius as a function of the chord in relation to the trigonometric partitions and take the first derivative, we will obtain the chord as a function of the trigonometric partitions.

The cosecant function is greater than the secant function; therefore, the cotangent function is greater than the tangent function,it is depending on which angle is being used .

Using the trigonometric partitions equation of the chord: $r_0(\theta) = r_n * e^{-2^{p+2} * \sin^2(\theta/2^{p+1})}$ and substituting it into the circle equation $x^2 + y^2 = r_n^2$ with $y = 1$ and $x = cot(\theta/2^p)$, since $cot^2(\theta/2^p) + 1 = csc^2(\theta/2^p)$, therefore we obtain $r_0(\theta) = csc(\theta/2^p) * e^{-2^{p+2} * \sin^2(\theta/2^{p+1})}$.

If we have the radius in trigonometric spiral form, as a function of the trigonometric partitions;

$r_n(\theta) = 0.5 * csc(\theta/2^p)$ and we substitute it into the final radius equation as a function of the trigonometric partitions;

$r_0(\theta) = r_n * e^{-2^{p+2} * \sin^2(\theta/2^{p+1})}$, we obtain $r_0(\theta) = 0.5 * csc(\theta/2^p) * e^{-2^{p+2} * \sin^2(\theta/2^{p+1})}$, and by generating the two plots of these two equations we will observe the importance of the value 0.5 and the cosecant function in the size of the graphs. We can do this same process with the trigonometric spiral equations without trigonometric partitions, obtaining two equations for their initial radii. In this way, we observe how a wavelet is born in relation to its initial radius, independent of the final radius.

This same process can be applied to the rest of the trigonometric partitions equations of the circular sector,the apothem, the arrow, the growth of the radius, or the radius equal to the sum of the apothem plus the arrow.

Figure 1.12: General Equation of the Trigonometric Partitions of the Chord.

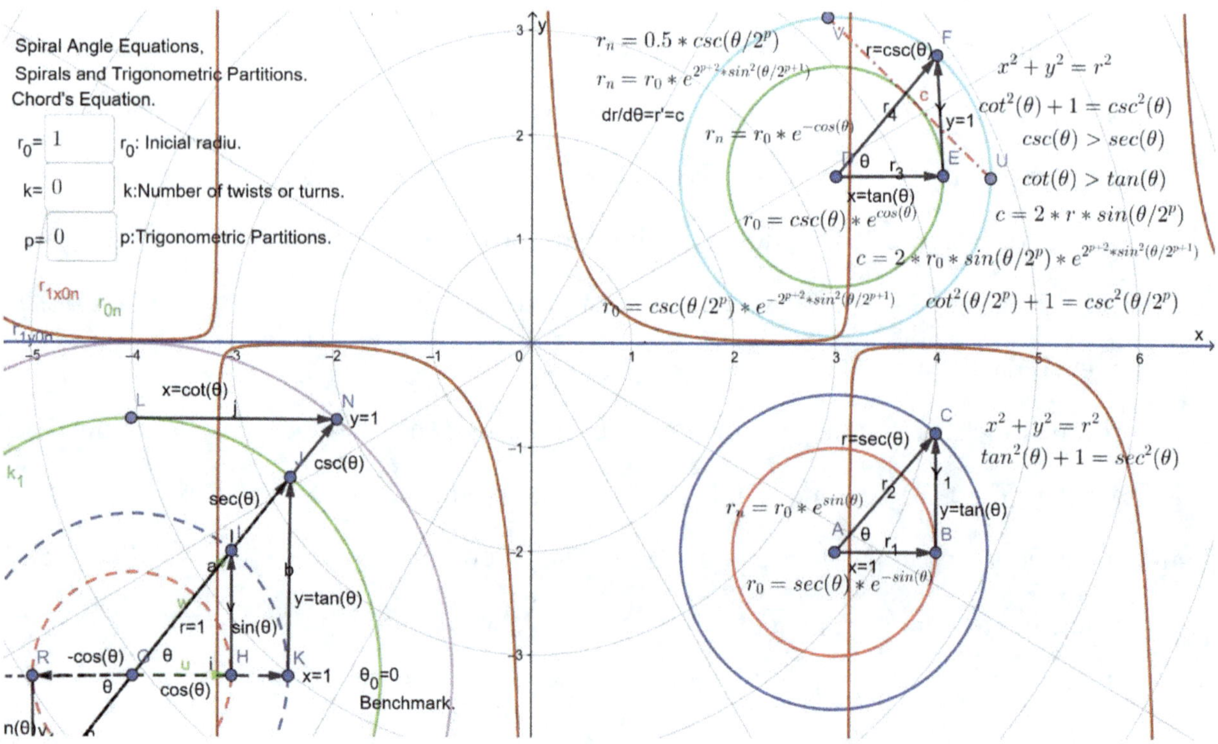

Note:In this graph, it can be observed how wavelets was born with an initial radius independent of their final radius, through the equations of trigonometric partitions.

Figure 1.13: General Equation of the Trigonometric Partitions of the Chord.

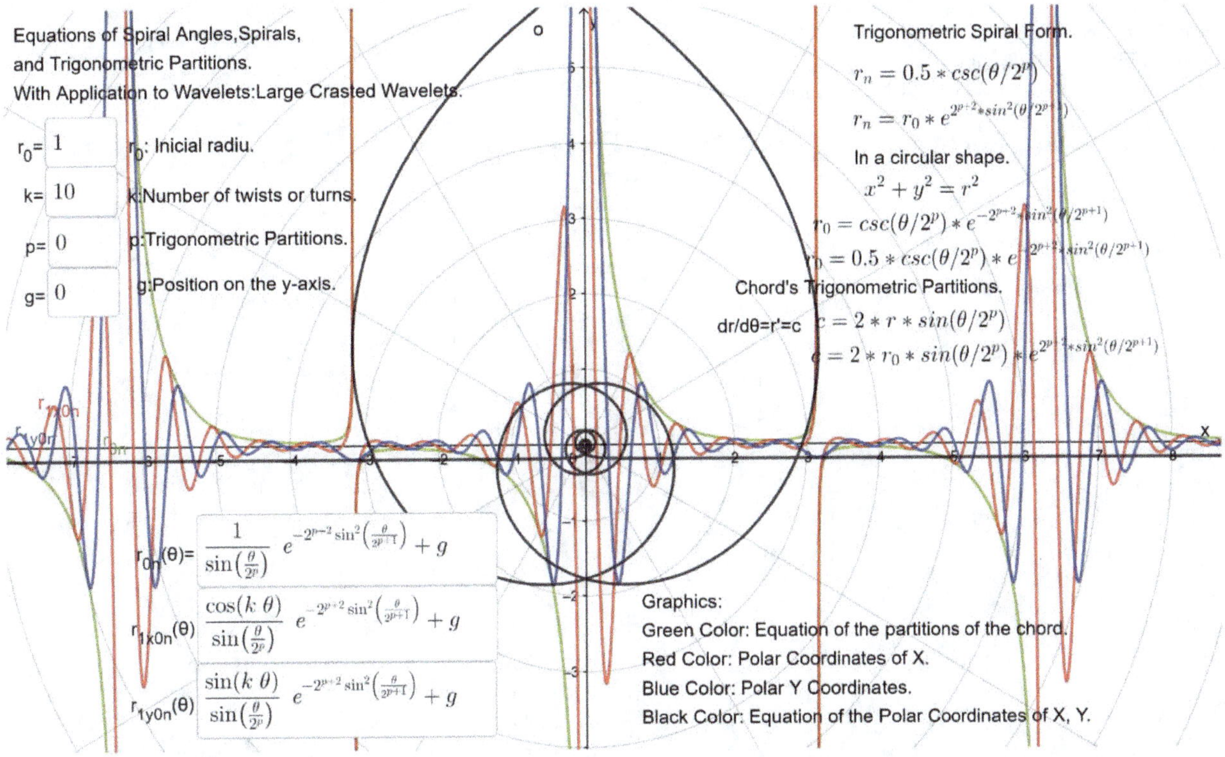

Note:In this graph, it can be seen how the wavelets was born with an initial radius independent of their final radius, using the equation of the trigonometric partitions of the chord.

Figure 1.14: General Equation of the Trigonometric Partitions of the Chord.

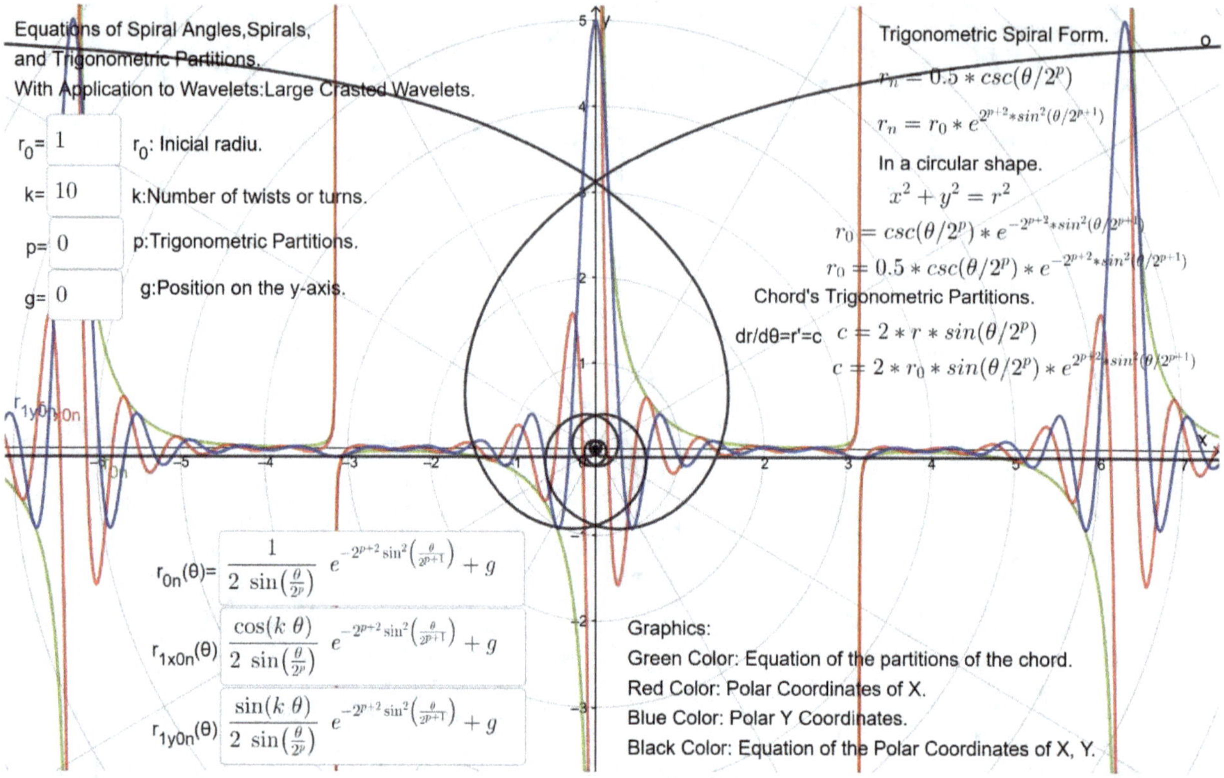

Note:In this graph, it can be seen how the wavelets appear with an initial radius independent of their final radius, using the equation of the trigonometric partitions of the chord with 0.5.

The case of the apothem is slightly different, since its radius is related to the secant of the angle. Using the trigonometric partitions equation of the apothem:

$r_0(\theta) = r_n * e^{-2^P * \sin(\theta/2^P)}$ and substituting it into the circle equation $x^2 + y^2 = r_n^2$ with $x = 1$ and $y = tan(\theta/2^p)$, since $tan^2(\theta/2^p) + 1 = sec^2(\theta/2^p)$, therefore we obtain

$r_0(\theta) = sec(\theta/2^p) * e^{-2^P * \sin(\theta/2^P)}$.

Figure 1.15: General Equation of the Trigonometric Partitions of the Apothem.

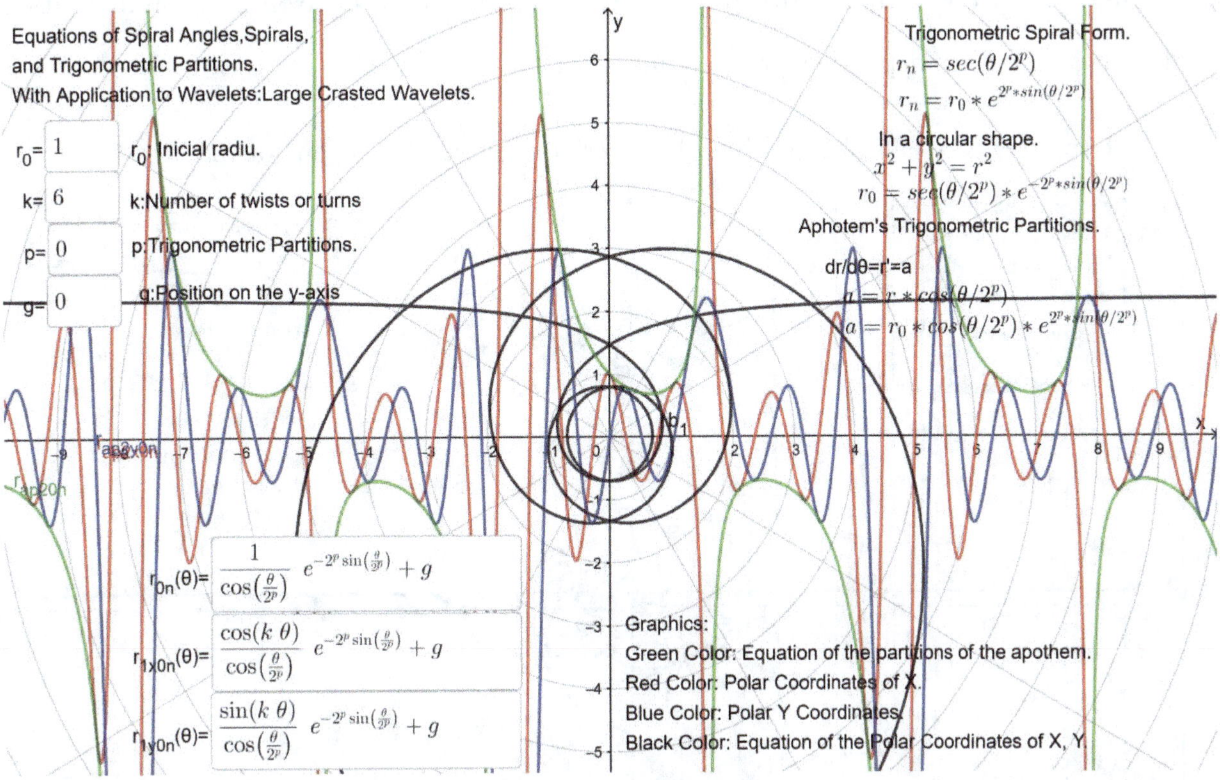

Note: In this graph, it can be observed how the wavelets start with an initial radius independent of their final radius, using the equation of the trigonometric partitions of the apothem.

Using the trigonometric partitions equation of the arrow:

$r_0(\theta) = r_n * e^{2^P * \sin(\theta/2^P) - \theta}$ and substituting it into the circle equation $x^2 + y^2 = r_n^2$ with $y = 1$ and $x = cot(\theta/2^p)$, since $cot^2(\theta/2^p) + 1 = csc^2(\theta/2^p)$, therefore we obtain;

$r_0(\theta) = csc(\theta/2^p) * e^{2^P * \sin(\theta/2^P) - \theta}$.

If we have the radius in trigonometric spiral form, as a function of the trigonometric partitions;

$r_n(\theta) = 0.5 * csc^2(\theta/2^{p+1})$ and we substitute it into the final radius equation as a function of the trigonometric partitions;

$r_0(\theta) = r_n * e^{2^p * \sin(\theta/2^p) - \theta}$, we will obtain $r_0(\theta) = 0.5 * csc^2(\theta/2^{p+1}) * e^{2^p * \sin(\theta/2^p) - \theta}$, and by generating the two plots of these two equations we will observe the importance of the value 0.5 and the cosecant function in the size of the graphs. We can carry out this same process with the trigonometric spiral equations without trigonometric partitions, obtaining two equations for their initial radii. In this way, we observe how a wavelet is born in relation to its initial radius, independent of the final radius.

Figure 1.16: General Equation of the Trigonometric Partitions of the arrow.

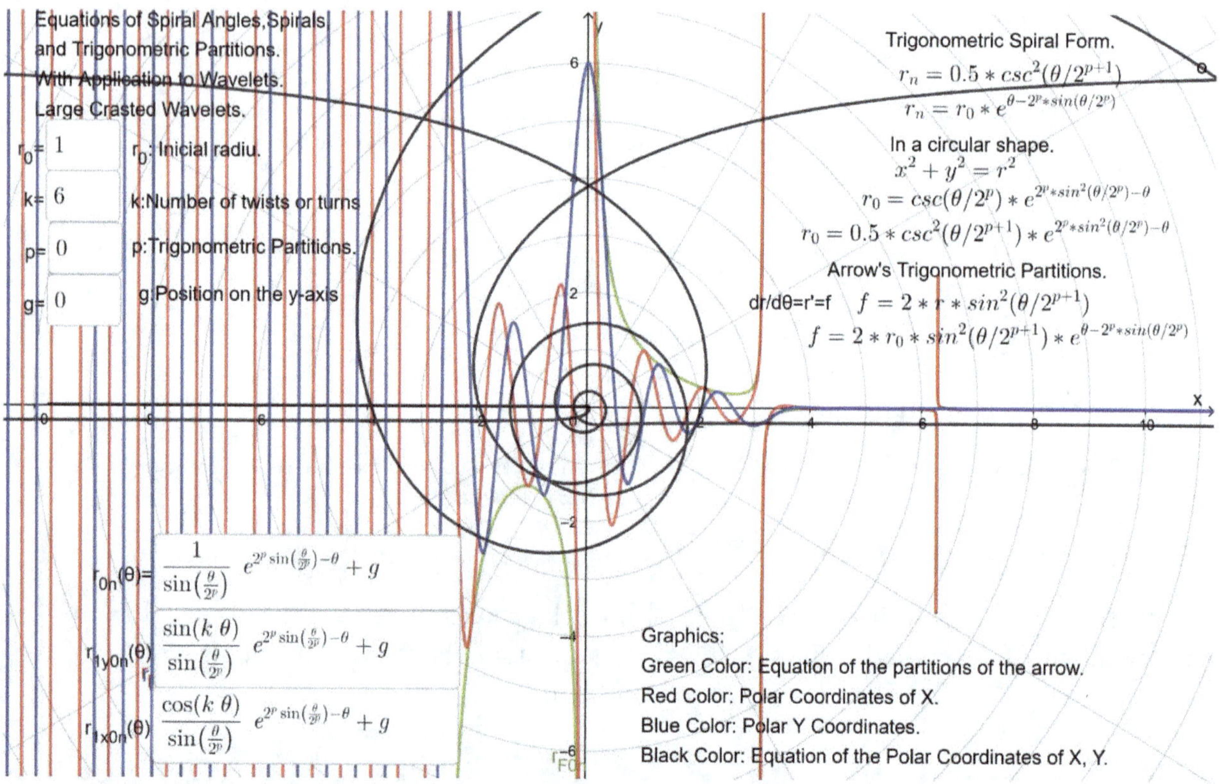

Note:In this graph, it can be seen how the wavelets emerge with an initial radius independent of their final radius, using the equation of the trigonometric partitions of the arrow.

Using the trigonometric partitions equation of radius growth:

$r_0(\theta) = r_n * e^{2^{p-1} * \sin^2(\theta/2^{p-1}) - \theta}$ and substituting it into the circle equation $x^2 + y^2 = r_n^2$ with $y = 1$ and $x = cot(\theta/2^p)$, since $cot^2(\theta/2^p) + 1 = csc^2(\theta/2^p)$, therefore we obtain

$r_0(\theta) = csc(\theta/2^p) * e^{2^{p-1} * \sin(\theta/2^{p-1}) - \theta}$.

If we have the radius in trigonometric spiral form, as a function of the trigonometric partitions;

$r_n(\theta) = 0.5 * csc^2(\theta/2^p)$ and we substitute it into the final radius equation as a function of the trigonometric partitions;

$r_0(\theta) = r_n * e^{-2^{p+2} * \sin^2(\theta/2^{p+1})}$, we obtain $r_0(\theta) = 0.5 * csc^2(\theta/2^p) * e^{2^p * \sin(\theta/2^p) - \theta}$, and by generating the two plots of these two equations we will observe the importance of the value 0.5 and the cosecant function in the size of the graphs. We can carry out this same process with the trigonometric spiral equations without trigonometric partitions, obtaining two equations for their initial radii. In this way, we observe how a wavelet is born in relation to its initial radius, independent of the final radius.

Figure 1.17: General Equation of the Trigonometric Partitions of radius growth.

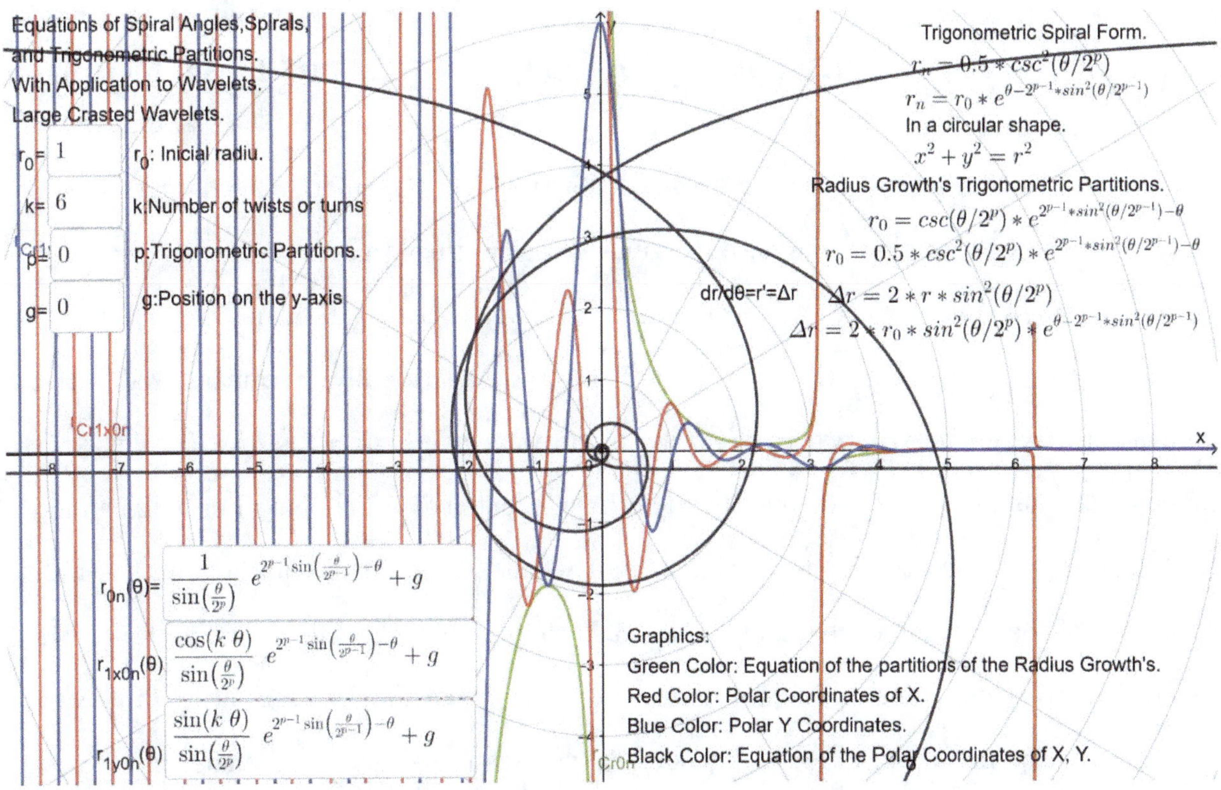

Note:In this graph, it can be observed how the wavelets are formed with an initial radius independent of their final radius, using the equation of the trigonometric partitions of radius growth.

Chapter 2

Wavelets With Trigonometric Partitions Of The Circle And Their Transforms.

2.1 Wavelets Transforms Of Trigonometric Partitions.

To perform the transforms of the trigonometric partitions of the circle we are going to require the identities of the angle as they are; Angular velocity times time: $\theta = \omega * t$,
Frequency times time: $\theta = 2*\pi*f*t$, or number of waves times position and their subtraction from angular velocity times time: $\theta = k * x - \omega * t$, Additionally, it can be used with its equivalent of the number of waves as a function of wavelength: $k = 2 * \pi/\lambda$. If we use these angle equations in each of the equations for the elements of a circle based on trigonometric partitions, we can create wavelets. Similarly, if we use these angle equations in the radius equations of the trigonometric partitions for each element of the circle, we will obtain a great diversity of wavelets. This technique offers us the ease of creating wavelets in two dimensions, creating their polar coordinates, and generating surface plots in three dimensions.
When analyzing each variable of each equation of the trigonometric partitions for generating wavelets, we can observe the effect that each one of them or their combination produces in the wavelet graphs. For example, the variable **k** causes the oscillations of the wavelet to

multiply; in other cases, it causes the crest of the wavelet to increase in size. The variable **p**, when performing positive cuts, stretches the oscillations of the wavelet, or compresses them when the cut is negative. The variable **t** causes the wave to shift longitudinally along the **x**-axis, or multiplies the number of longitudinal waves, all depending on the type of equation being used. The variable x_0 causes the wave to shift longitudinally along the **x**-axis, without altering the wave shape. The angular velocity ω, if positive, causes the wave to move to the right, and if negative, it moves to the left, without changing the wavelet. The frequency **f** compresses the wavelet more as it increases. The wavelength λ causes the oscillations to either multiply or decrease.

2.2 Wavelets Transforms Of The Trigonometric Partitions Of Large Crest.

In this study of large crest wavelets, it is based on the trigonometric partitions of the chord, and in the same way we can do this for the rest of the trigonometric partitions such as the circular sector, the apothem, the arrow,and the growth of the radius, etc.
The equations of trigonometric partitions can be applied in classical physics as well as in quantum physics. The first derivative of the radius as a function of time allows us to obtain the radial velocity of the radius, and if we decompose it into its **x** and **y** components, we can obtain the radial and tangential velocity; likewise, its second derivative can be obtained to compute the radial acceleration of the radius and obtain its **x** and **y** components to compute the radial and tangential accelerations. These second derivative processes are large. In most of these cases, the power **n** equal to one will be used for their transformations.
If we perform the substitution of the equivalent equation of the angle as a function of the angular velocity with respect to time, in the general equation of the trigonometric partitions of the chord, we obtain: $\theta = \omega * t$. With the power n=1
$r_0(\theta) = (0.5 * csc(\theta))^n * e^{-2^{p+2} * n * \sin^2(\theta/2^{p+1})}$.The envelope equation of the wavelet.
$r_{0x}(\theta) = (0.5 * csc(\theta))^n * \cos^n(k * \theta) * e^{-2^{p+2} * n * \sin^2(\theta/2^{p+1})}$.The equation of the **x**-coordinate of the wavelet.
$r_{0y}(\theta) = (0.5 * csc(\theta))^n * sin^n(k * \theta) * e^{-2^{p+2} * n * sin^2(\theta/2^{p+1})}$.The equation of the **y**-coordinate of the wavelet.
These equations of the radius can be written as a function of angular velocity and time, generating wavelets. If we wish to move it to any position on the **y**-axis, it can be done with the constant **g** of the trigonometric partition equation of the radius.

Figure 2.1: General Equation of the Trigonometric Partitions of the Chord.

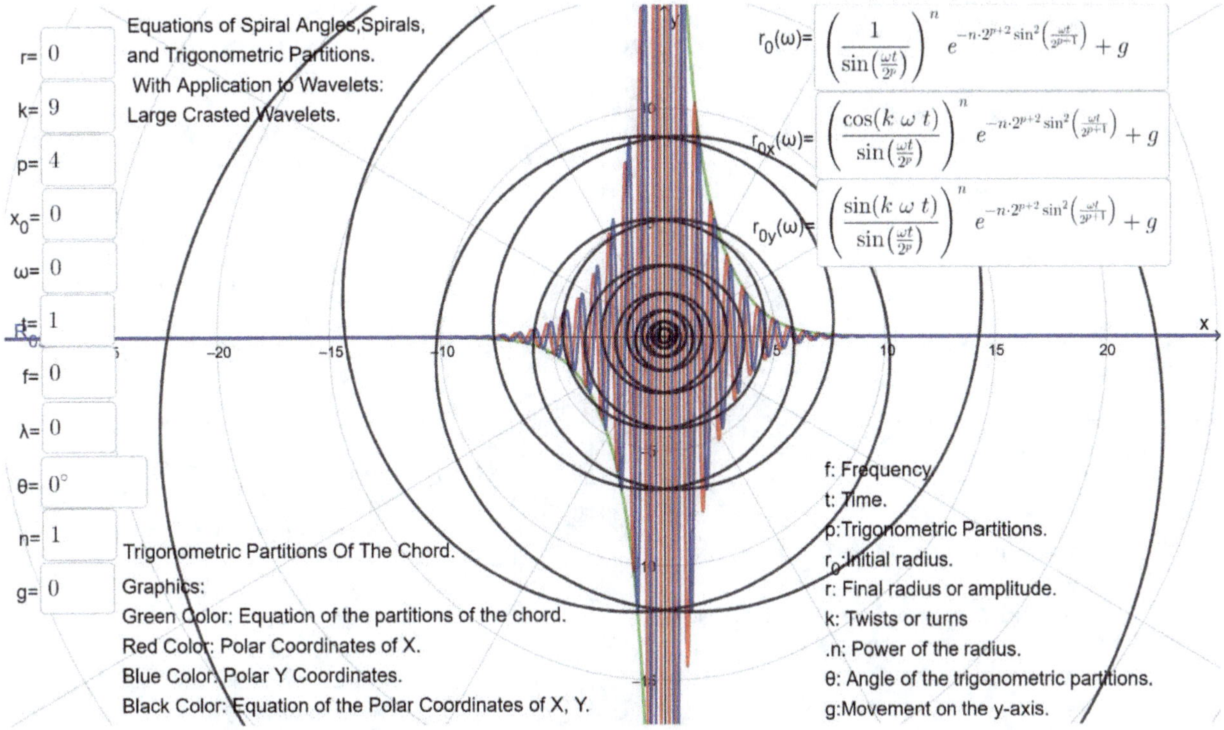

Note:In this graph we observe how large crest wavelets are generated by the powers n with the initial radius equations as a function of the angular velocity.

Furthermore, we can do the same if we substitute; $\theta = 2 * \pi * \mathrm{f} * \mathrm{t}$, with n=1 in the equations of the trigonometric partitions of the chord of the radius as a function of the angle:

$\mathrm{r}_0(\theta) = (0.5 * \csc(\theta))^{\mathrm{n}} * \mathrm{e}^{-2^{p+2} * \mathrm{n} * \sin^2(\theta/2^{p+1})}$.

$\mathrm{r}_{0\mathrm{x}}(\theta) = (0.5 * \csc(\theta))^{\mathrm{n}} * \cos^{\mathrm{n}}(\mathrm{k} * \theta) * \mathrm{e}^{-2^{\mathrm{p}+2} * \mathrm{n} * \sin^2(\theta/2^{\mathrm{p}+1})}$.

$\mathrm{r}_{0\mathrm{y}}(\theta) = (0.5 * \csc(\theta))^{\mathrm{n}} * \sin^{\mathrm{n}}(\mathrm{k} * \theta) * \mathrm{e}^{-2^{\mathrm{p}+2} * \mathrm{n} * \sin^2(\theta/2^{\mathrm{p}+1})}$.

These equations of the radius can be written as a function of frequency and time, generating wavelets. If we wish to move it to any position on the **y**-axis, it can be done with the constant **g** of the trigonometric partition equation of the radius.

Figure 2.2: General Equation of the Trigonometric Partitions of the Chord.

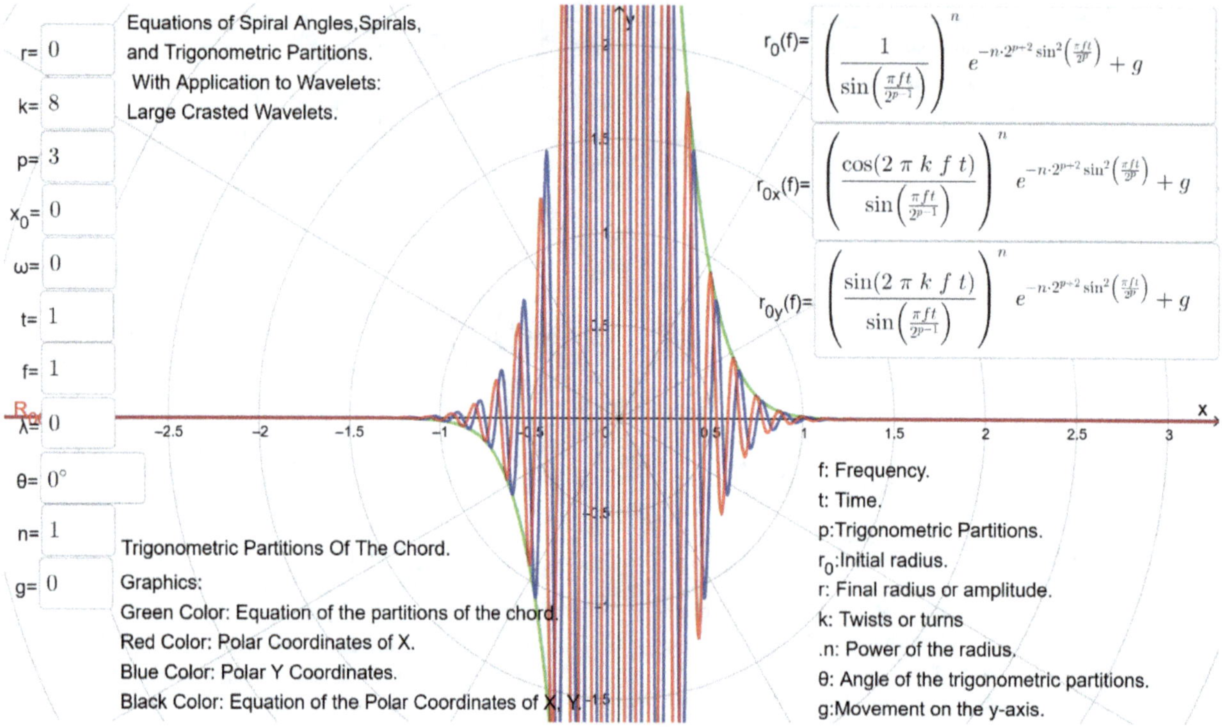

Note:In this graph we observe how large crest wavelets are generated by the powers n=1 with the initial radius equations as a function of time.

If we continue doing the same with the angle equation $\theta = k * x_0 - \omega * t$, with n=1, and substitute this equation into the function of the trigonometric partitions of the chord in the equation of the radius as a function of the angle, we obtain:

$r_0(\theta) = (0.5 * csc(\theta))^n * e^{-2^{p+2} * n * \sin^2(\theta/2^{p+1})}$.

$r_{0x}(\theta) = (0.5 * csc(\theta))^n * \cos^n(k * \theta) * e^{-2^{p+2} * n * \sin^2(\theta/2^{p+1})}$.

$r_{0y}(\theta) = (0.5 * csc(\theta))^n * \sin^n(k * \theta) * e^{-2^{p+2} * n * \sin^2(\theta/2^{p+1})}$.

These equations of the radius can be written as a function of the number of waves, the position on the x-axis, the angular velocity, and time; generating wavelets. If we wish to move it to any position on the **y**-axis, it can be done with the constant **g** of the trigonometric partition equation of the radius.

Figure 2.3: General Equation of the Trigonometric Partitions of the Chord.

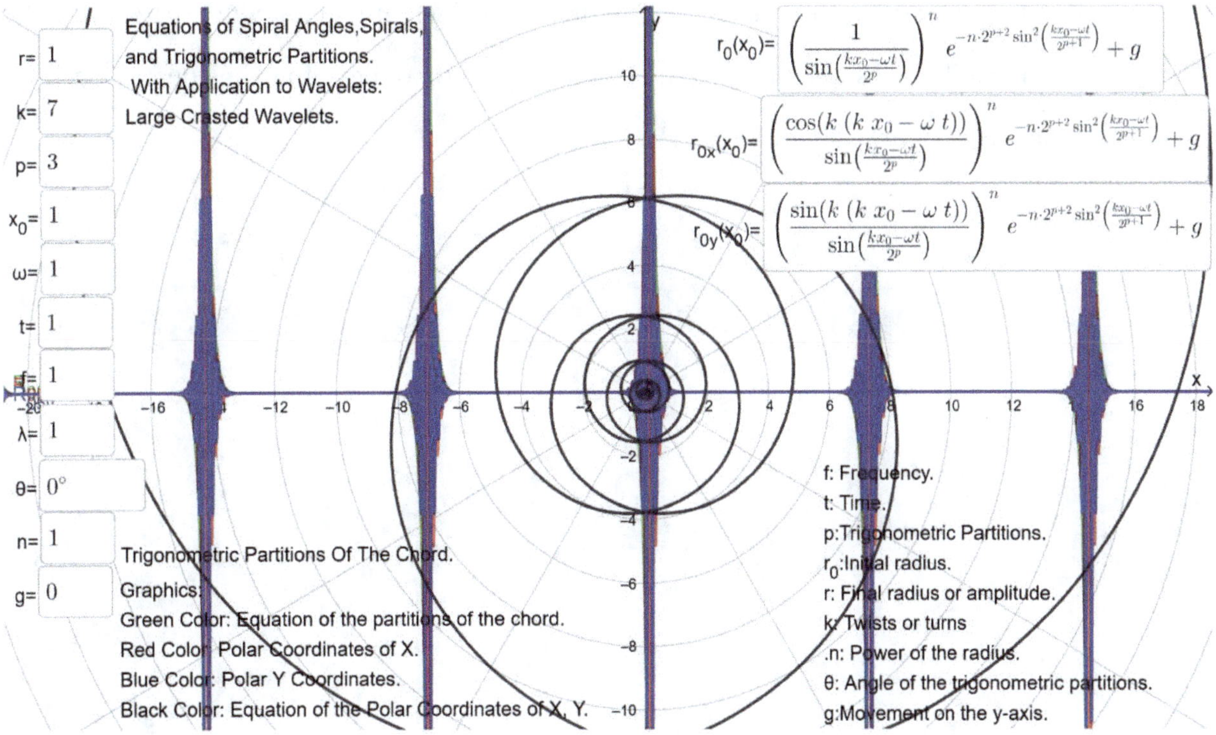

Note: In this graph we observe how large crest wavelets are generated by the powers n=1 with the initial radius equations as a function of the position on the **x**-axis.

Finally, if we continue doing the same with the angle equation $\theta = 2 * \pi * x_0/\lambda - \omega * t$, with $k = 2 * \pi/\lambda$, with n=1, and substitute this equation into the function of the trigonometric partitions of the chord in the equation of the radius as a function of the angle, we obtain:

$r_0(\theta) = (0.5 * csc(\theta))^n * e^{-2^{p+2}*n*\sin^2(\theta/2^{p+1})}$.

$r_{0x}(\theta) = (0.5 * csc(\theta))^n * \cos^n(k * \theta) * e^{-2^{p+2}*n*\sin^2(\theta/2^{p+1})}$.

$r_{0y}(\theta) = (0.5 * csc(\theta))^n * \sin^n(k * \theta) * e^{-2^{p+2}*n*\sin^2(\theta/2^{p+1})}$.

These equations of the radius can be written as a function of the wavelength, the position on the x-axis, the angular velocity, and time; generating wavelets. If we wish to move it to any position on the **y**-axis, it can be done with the constant **g** of the trigonometric partition equation of the radius. These radius equations can be expressed as a function of the wavelength, the position on the x-axis, the angular velocity, and time; generating wavelets. If we wish to move it to any position on the **y**-axis, it can be done using the constant **g** from the trigonometric partitions equation of the radius.

Figure 2.4: General Equation of the Trigonometric Partitions of the Chord.

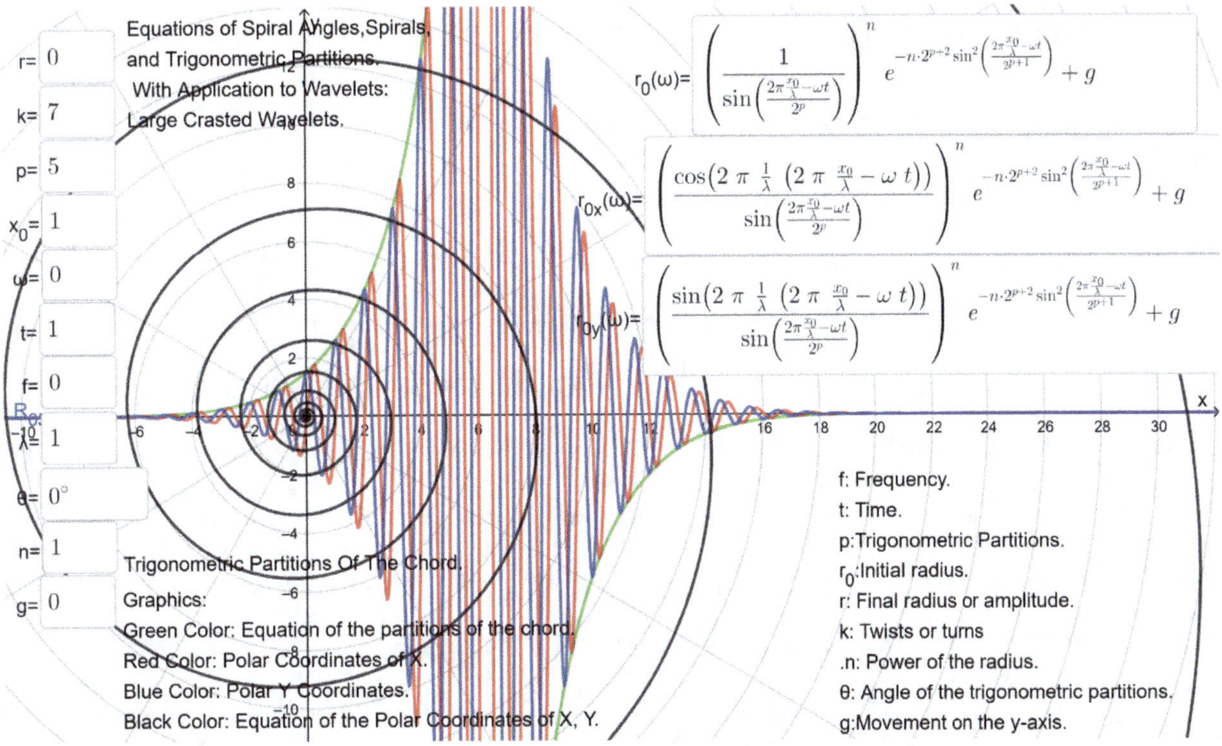

Note: In this graph, you can see how the large crest wavelets are generated by the powers n=1 using the equations of the initial radius as a function of the angular velocity ω.

2.3 Wavelets Transforms Of The Trigonometric Partitions Of The Circular Sector.

If we substitute the equivalent equation of the angle as a function of angular velocity for time, in the general equation of the trigonometric partitions of the circular sector we will obtain: $\theta = \omega * t$.

$r_0(\theta) = r^n * e^{-2^{2*p+1}*n*\sin^2(\theta/2^{p+1})}$. The envelope equation of the wavelet

$r_{0x}(\theta) = r^n * \cos^n(k*\theta) * e^{-2^{2*p+1}*n*\sin^2(\theta/2^{p+1})}$. The equation of the **x**-coordinate of the wavelet.

$r_{0y}(\theta) = r^n * \sin^n(k*\theta) * e^{-2^{2*p+1}*n*\sin^2(\theta/2^{p+1})}$. The equation of the **x**-coordinate of the wavelet.

These radius equations can be written as a function of angular velocity and time, generating wavelets. If we want to move it to any position on the **y** axis, we just have to add a constant **g** to the equation of the trigonometric partitions of the radius.

Figure 2.5: General Equation of Trigonometric Partitions of the Circular Sector.

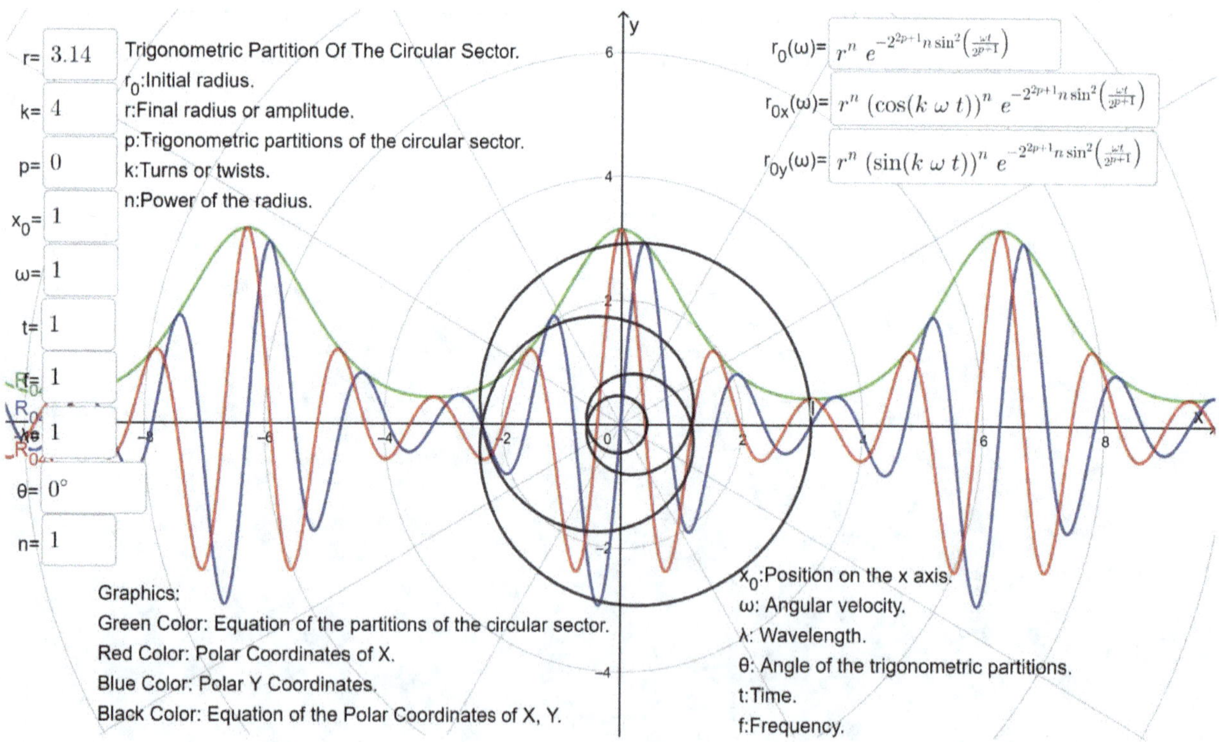

Note: This graph shows how the wavelets are generated by the powers n with the equations of the initial radius as a function of the angular velocity.

Also, we can do the same if we substitute; $\theta = 2 * \pi * f * t$ in the equations of the trigonometric partitions of the circular sector of the radius as a function of the angle:

$r_0(\theta) = r^n * e^{-2^{2*p+1}*n*\sin^2(\theta/2^{p+1})}$.

$r_{0x}(\theta) = r^n * \cos^n(k * \theta) * e^{-2^{2*p+1}*n*\sin^2(\theta/2^{p+1})}$.

$r_{0y}(\theta) = r^n * \sin^n(k * \theta) * e^{-2^{2*p+1}*n*\sin^2(\theta/2^{p+1})}$.

These radius equations can be written as a function of frequency and time, generating wavelets. If we want to move it to any position on the **y** axis, we just have to add a constant **g** to the equation of the trigonometric partitions of the radius.

With the first derivatives of these equations of the initial radius as a function of the trigonometric partitions **p** we can create scalograms.

Figure 2.6: General Equation Of Trigonometric Partitions Of The Circular Sector.

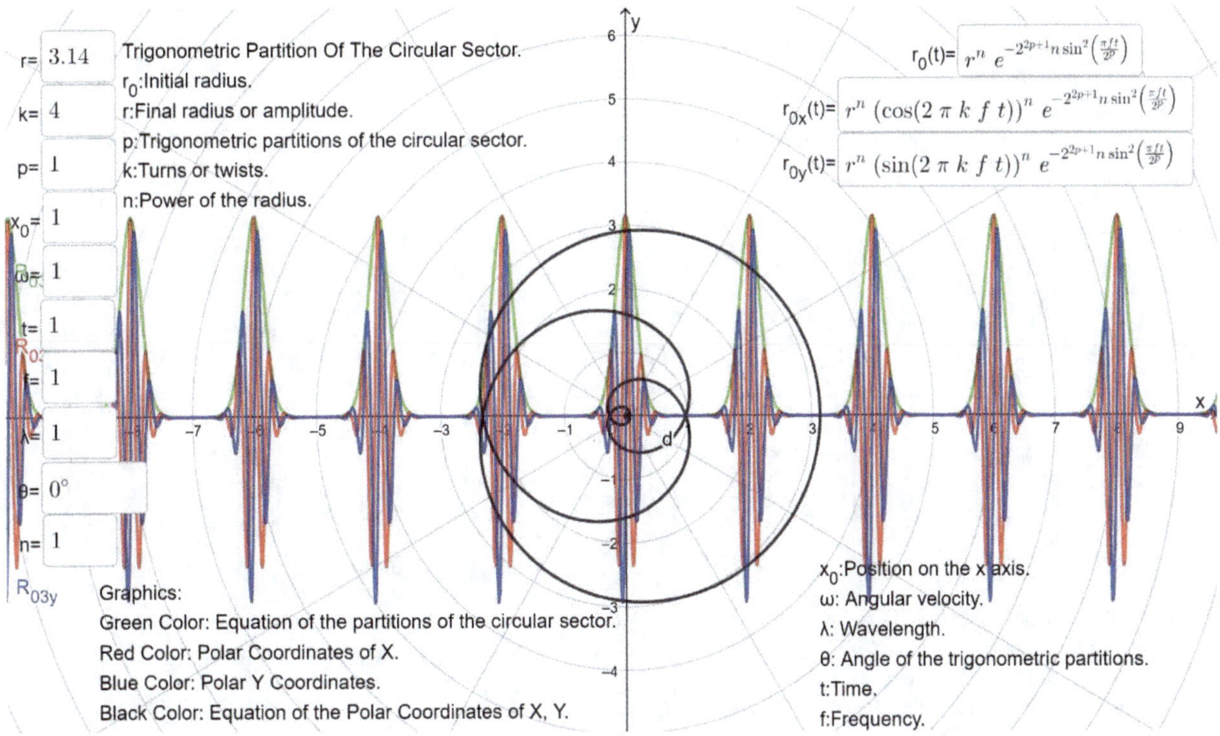

Note:This graph shows how the wavelets are generated by the powers n=1 with the equations of the initial radius as a function of time.

If we continue doing the same with the angle equation $\theta = k * x_0 - \omega * t$, if we substitute this equation in the function of the trigonometric partitions of the circular sector in the equation of the radius as a function of the angle we will obtain:

$r_0(\theta) = r^n * e^{-2^{2*p+1}*n*\sin^2(\theta/2^{p+1})}$.

$r_{0x}(\theta) = r^n * \cos^n(k * \theta) * e^{-2^{2*p+1}*n*\sin^2(\theta/2^{p+1})}$.

$r_{0y}(\theta) = r^n * \sin^n(k * \theta) * e^{-2^{2*p+1}*n*\sin^2(\theta/2^{p+1})}$.

These radius equations can be written as a function of the number of waves, the position on the x-axis, the angular velocity, and time; generating wavelets. If we want to move it to any position on the **y** axis, we only have to add a constant **g** to the equation of the trigonometric partitions of the radius.

Figure 2.7: General Equation Of Trigonometric Partitions of The Circular Sector.

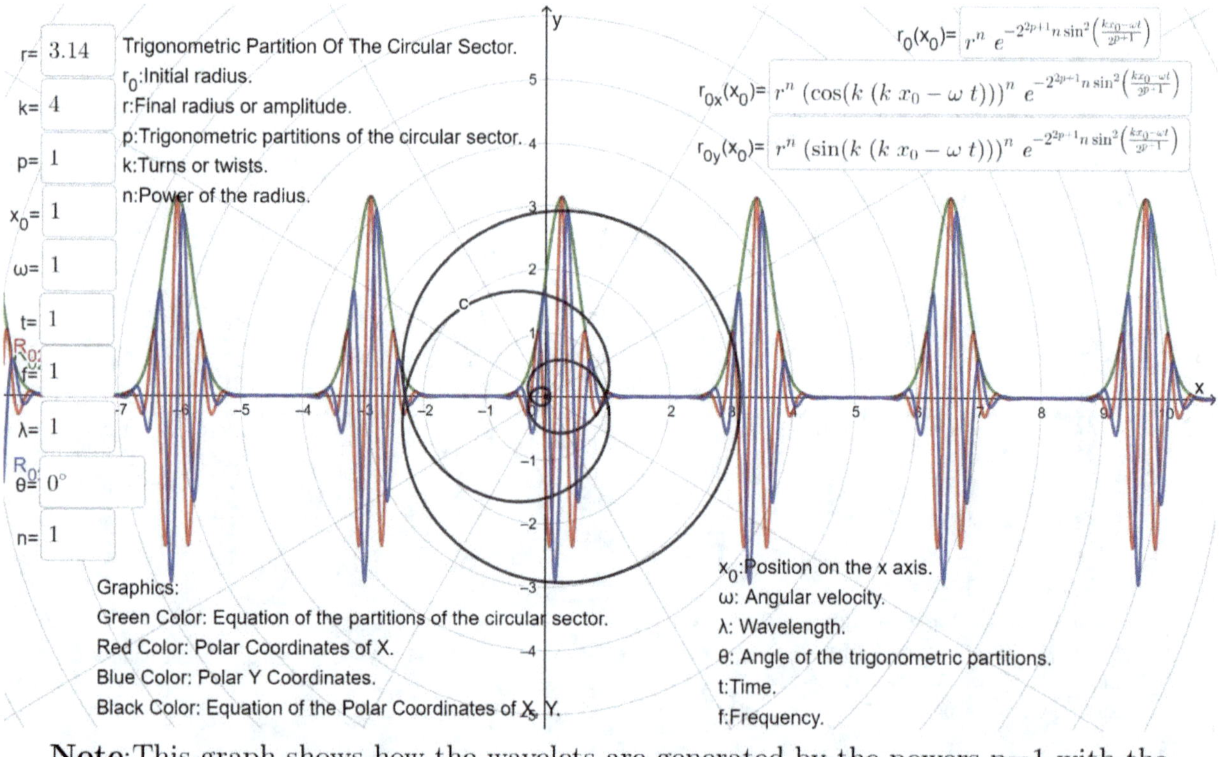

Note: This graph shows how the wavelets are generated by the powers n=1 with the equations of the initial radius as a function of the position on the **x** axis.

Finally, if we continue doing the same with the equation of the angle $\theta = 2*\pi*x_0/\lambda - \omega*t$, with $k = 2*\pi/\lambda$, if we substitute these equations in the function of the trigonometric partitions of the circular sector in The equation of the radius as a function of the angle we will obtain:

$r_0(\theta) = r^n * e^{-2^{2*p+1}*n*\sin^2(\theta/2^{p+1})}$.

$r_{0x}(\theta) = r^n * \cos^n(k*\theta) * e^{-2^{2*p+1}*n*\sin^2(\theta/2^{p+1})}$.

$r_{0y}(\theta) = r^n * \sin^n(k*\theta) * e^{-2^{2*p+1}*n*\sin^2(\theta/2^{p+1})}$.

These radius equations can be written as a function of wavelength, position on the x-axis, angular velocity, and time; generating wavelets. If we want to move it to any position on the **y** axis, we just have to add a constant **g** to the equation of the trigonometric partitions of the radius.

Figure 2.8: General Equation Of Trigonometric Partitions Of The Circular Sector.

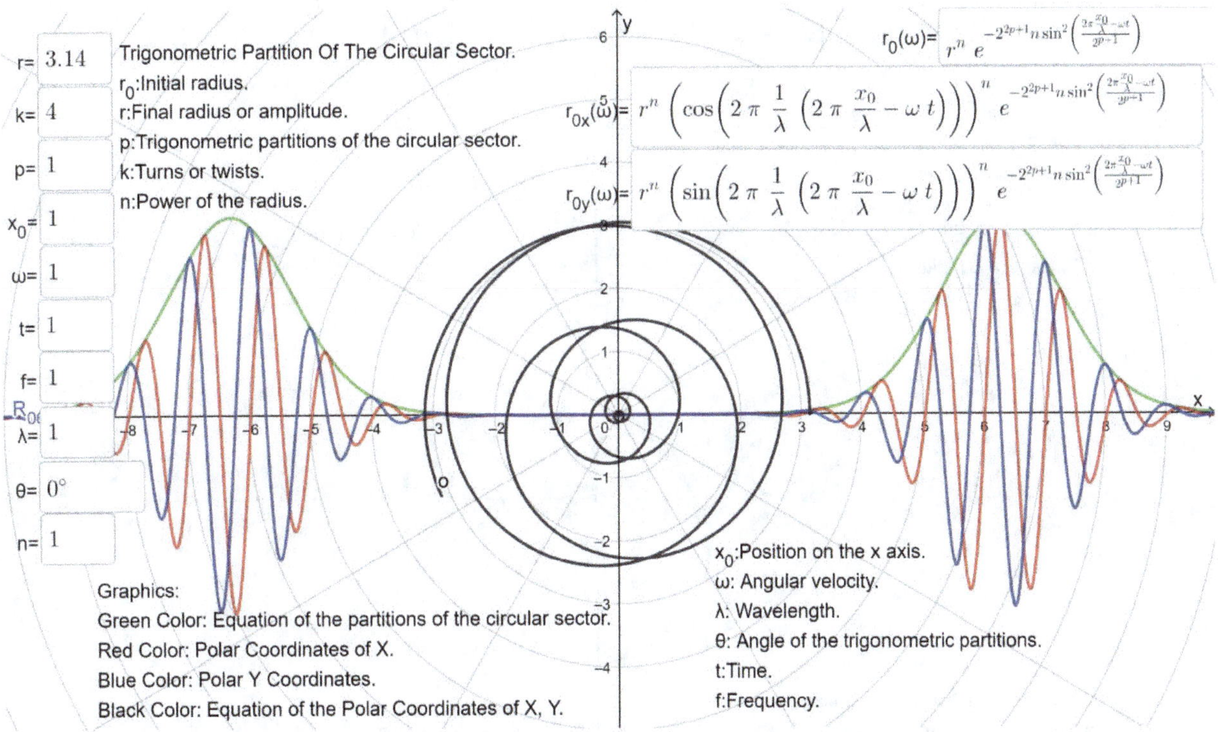

Note:This graph shows how the wavelets are generated by the powers n=1 with the equations of the initial radius as a function of the angular velocity ω.

2.4 Wavelets Transforms Of Trigonometric Partitions Of The Chord.

If we substitute the equivalent equation of the angle as a function of angular velocity for time, in the general equation of the trigonometric partitions of the chord we will obtain: $\theta = \omega * t$.

$r_0(\theta) = r^n * e^{-2^{p+2} * n * \sin^2(\theta/2^{p+1})}$.

$r_{0x}(\theta) = r^n * \cos^n(k * \theta) * e^{-2^{p+2} * n * \sin^2(\theta/2^{p+1})}$.

$r_{0y}(\theta) = r^n * \sin^n(k * \theta) * e^{-2^{p+2} * n * \sin^2(\theta/2^{p+1})}$.

These radius equations can be written as a function of angular velocity and time, generating wavelets. If we want to move it to any position on the **y** axis, we just have to add a constant **g** to the equation of the trigonometric partitions of the radius.

Figure 2.9: General Equation Of Trigonometric Partitions Of The Chord.

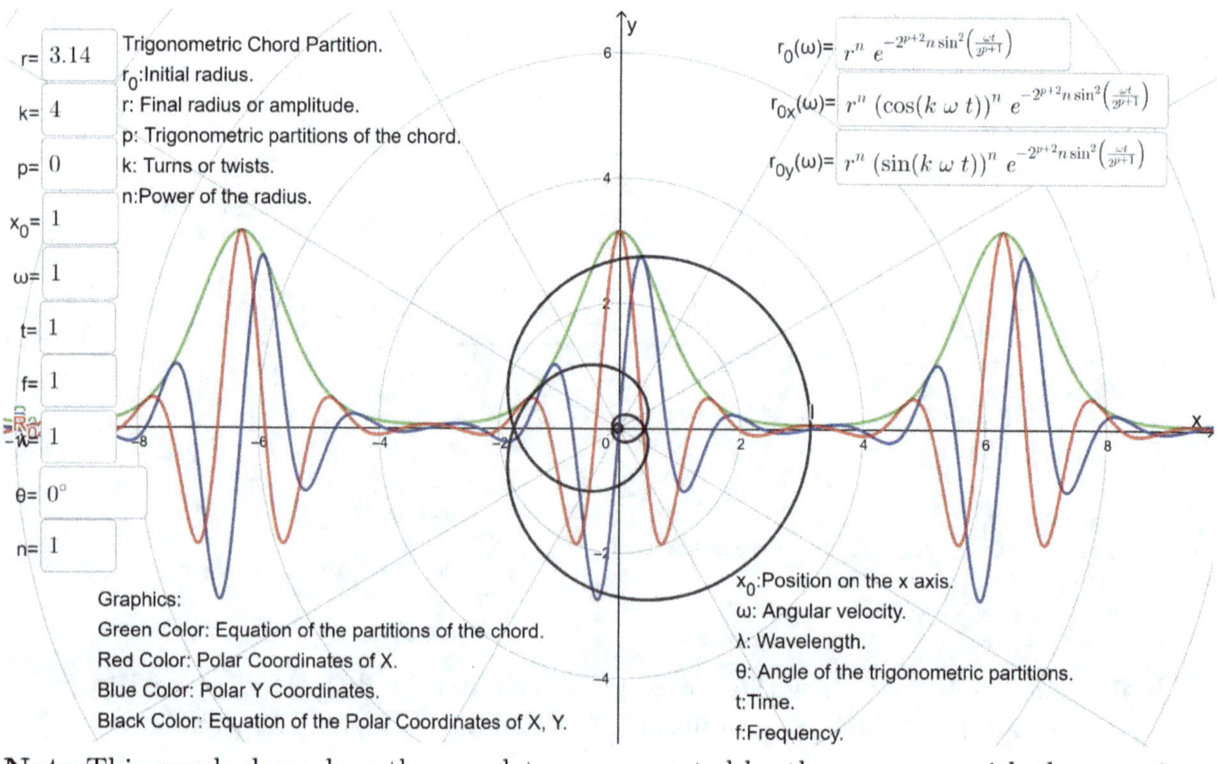

Note: This graph shows how the wavelets are generated by the powers n with the equations of the initial radius as a function of the angular velocity.

Also, we can do the same if we substitute; $\theta = 2 * \pi * f * t$ in the equations of the trigonometric partitions of the chord of the radius as a function of the angle:

$r_0(\theta) = r^n * e^{-2^{p+2} * n * \sin^2(\theta/2^{p+1})}$.

$r_{0x}(\theta) = r^n * \cos^n(k * \theta) * e^{-2^{p+2} * n * \sin^2(\theta/2^{p+1})}$.

$r_{0y}(\theta) = r^n * \sin^n(k * \theta) * e^{-2^{p+2} * n * \sin^2(\theta/2^{p+1})}$.

These radius equations can be written as a function of frequency and time, generating wavelets. If we want to move it to any position on the **y** axis, we just have to add a constant **g** to the equation of the trigonometric partitions of the radius.(José Mauricio Orellana Díaz, n.d.-a)

Figure 2.10: General Equation of Trigonometric Partitions of the Chord.

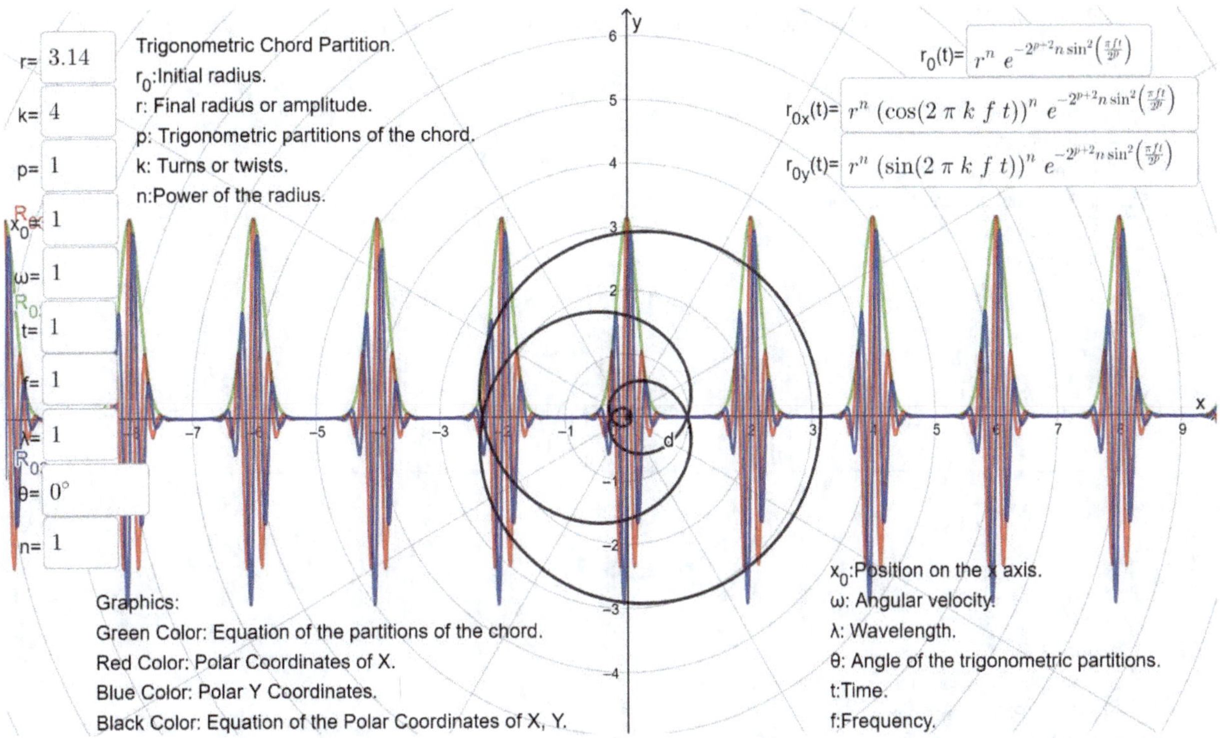

Note:This graph shows how the wavelets are generated by the powers n=1 with the equations of the initial radius as a function of time.

If we continue doing the same with the equation of the angle $\theta = k * x_0 - \omega * t$, if we substitute this equation in the function of the trigonometric partitions of the chord in the equation of the radius as a function of the angle we will obtain:

$r_0(\theta) = r^n * e^{-2^{p+2} * n * \sin^2(\theta/2^{p+1})}$.

$r_{0x}(\theta) = r^n * \cos^n(k * \theta) * e^{-2^{p+2} * n * \sin^2(\theta/2^{p+1})}$.

$r_{0y}(\theta) = r^n * \sin^n(k * \theta) * e^{-2^{p+2} * n * \sin^2(\theta/2^{p+1})}$.

These radius equations can be written as a function of the number of waves, the position on the x-axis, the angular velocity, and time; generating wavelets. If we want to move it to any position on the **y** axis, we only have to add a constant **g** to the equation of the trigonometric partitions of the radius.(José Mauricio Orellana Díaz, n.d.-b)

Figure 2.11: General Equation of Trigonometric Partitions of the Chord.

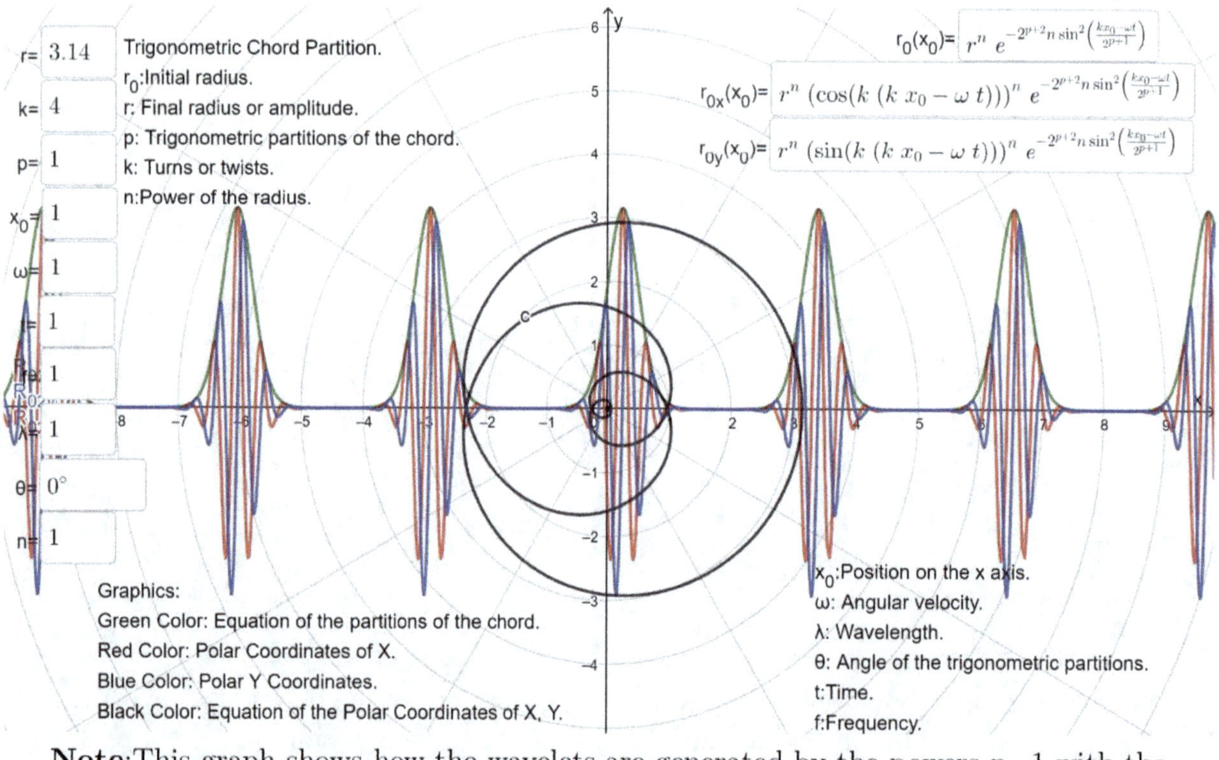

Note:This graph shows how the wavelets are generated by the powers n=1 with the equations of the initial radius as a function of the position on the **x** axis.

Finally, if we continue doing the same with the equation of the angle $\theta = 2 * \pi * x_0/\lambda - \omega * t$, with $k = 2 * \pi/\lambda$, if we substitute these equations in the function of the trigonometric partitions of the chord in the equation of the radius as a function of the angle we will obtain:

$r_0(\theta) = r^n * e^{-2^{p+2} * n * \sin^2(\theta/2^{p+1})}$.

$r_{0x}(\theta) = r^n * \cos^n(k * \theta) * e^{-2^{p+2} * n * \sin^2(\theta/2^{p+1})}$.

$r_{0y}(\theta) = r^n * \sin^n(k * \theta) * e^{-2^{p+2} * n * \sin^2(\theta/2^{p+1})}$.

These radius equations can be written as a function of wavelength, position on the x-axis, angular velocity, and time; generating wavelets. If we want to move it to any position on the **y** axis, we just have to add a constant **g** to the equation of the trigonometric partitions of the radius.

Figure 2.12: General Equation Of Trigonometric Partitions Of The Chord.

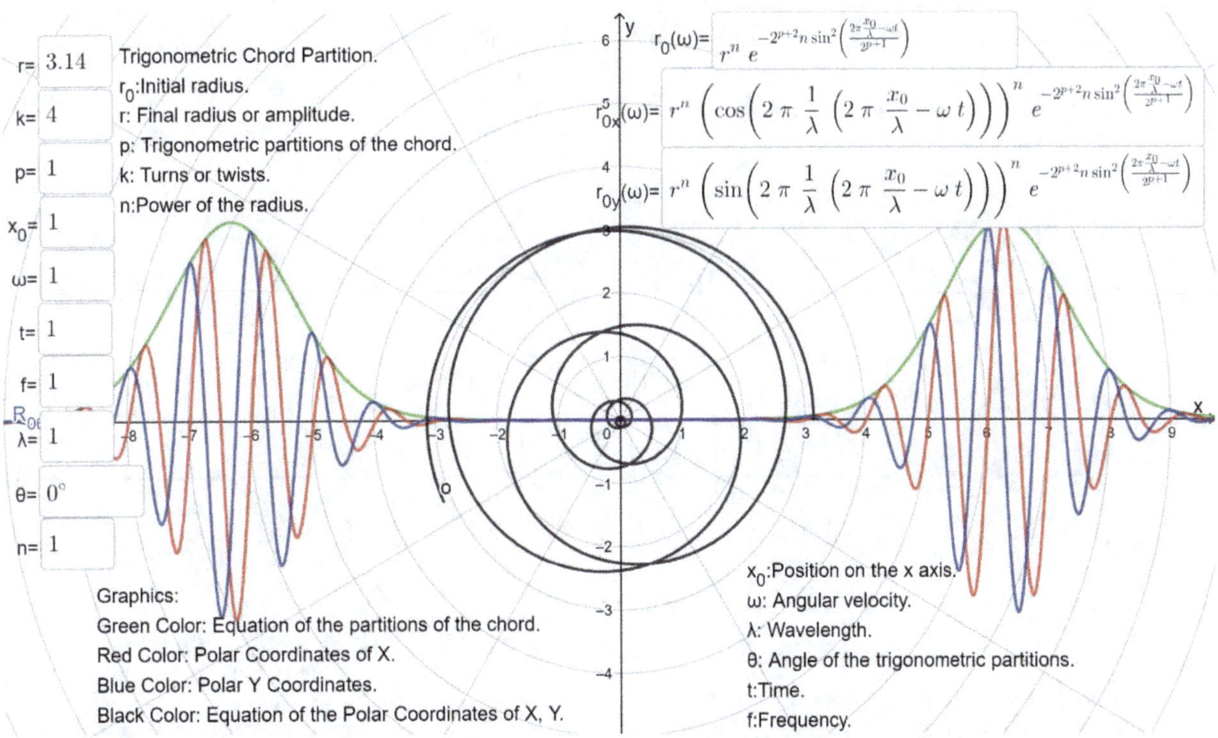

Note:This graph shows how the wavelets are generated by the powers n=1 with the equations of the initial radius as a function of the angular velocity ω.

2.5 Wavelets Transforms Of The Trigonometric Partitions Of The Apothem.

If we substitute the equivalent equation of the angle as a function of angular velocity for time, in the general equation of the trigonometric partitions of the apothem we will obtain:

$\theta = \omega * t.$

$r_0(\theta) = r^n * e^{-2^p*n*\sin(\theta/2^p)}.$

$r_{0x}(\theta) = r^n * \cos^n(k*\theta) * e^{-2^p*n*\sin(\theta/2^p)}.$

$r_{0y}(\theta) = r^n * \sin^n(k*\theta) * e^{-2^p*n*\sin(\theta/2^p)}.$

These radius equations can be written as a function of angular velocity and time, generating wavelets. If we want to move it to any position on the **y** axis, we just have to add a constant **g** to the equation of the trigonometric partitions of the radius.

Figure 2.13: General Equation Of Trigonometric Partitions of The Apothem.

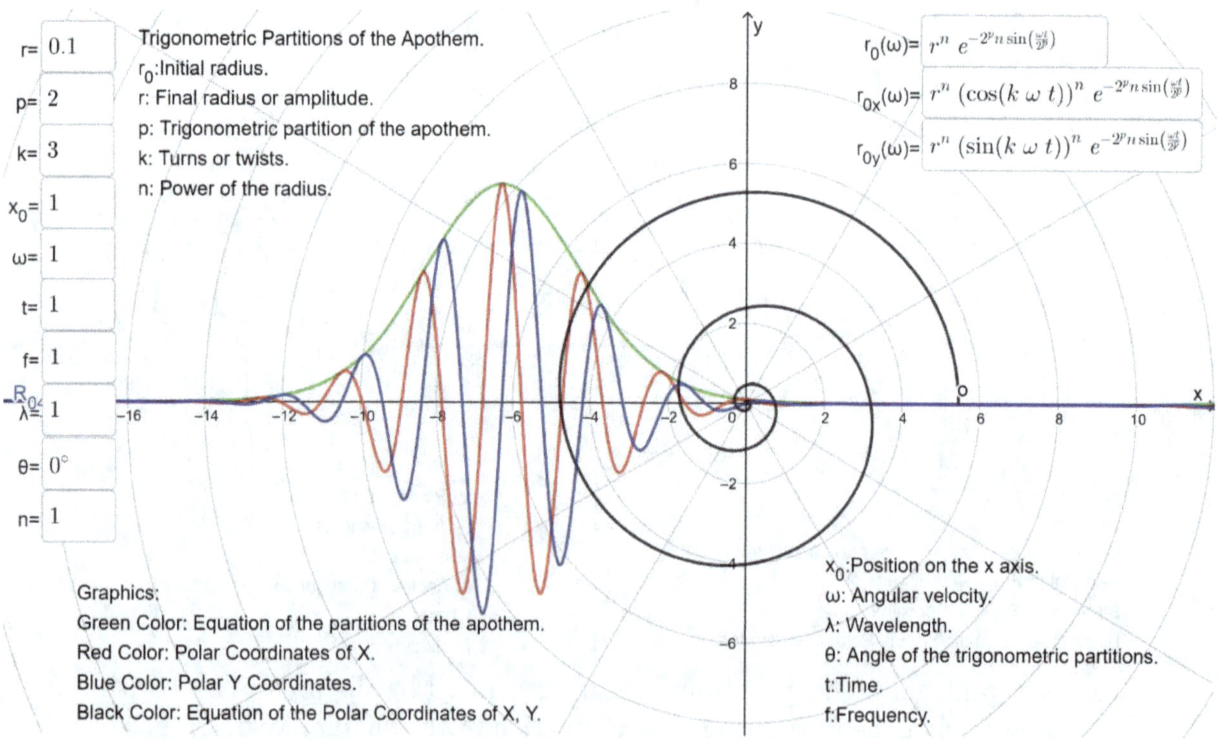

Note:This graph shows how the wavelets are generated by the powers n with the equations of the initial radius as a function of the angular velocity.

Also, we can do the same if we substitute; $\theta = 2 * \pi * f * t$ in the equations of the trigonometric partitions of the apothem of the radius as a function of the angle:

$r_0(\theta) = r^n * e^{-2^P * n * \sin(\theta/2^P)}$.

$r_{0x}(\theta) = r^n * \cos^n(k * \theta) * e^{-2^P * n * \sin(\theta/2^P)}$.

$r_{0y}(\theta) = r^n * \sin^n(k * \theta) * e^{-2^P * n * \sin(\theta/2^P)}$.

These radius equations can be written as a function of frequency and time, generating wavelets. If we want to move it to any position on the **y** axis, we just have to add a constant **g** to the equation of the trigonometric partitions of the radius.

Figure 2.14: General Equation Of Trigonometric Partitions Of The Apothem.

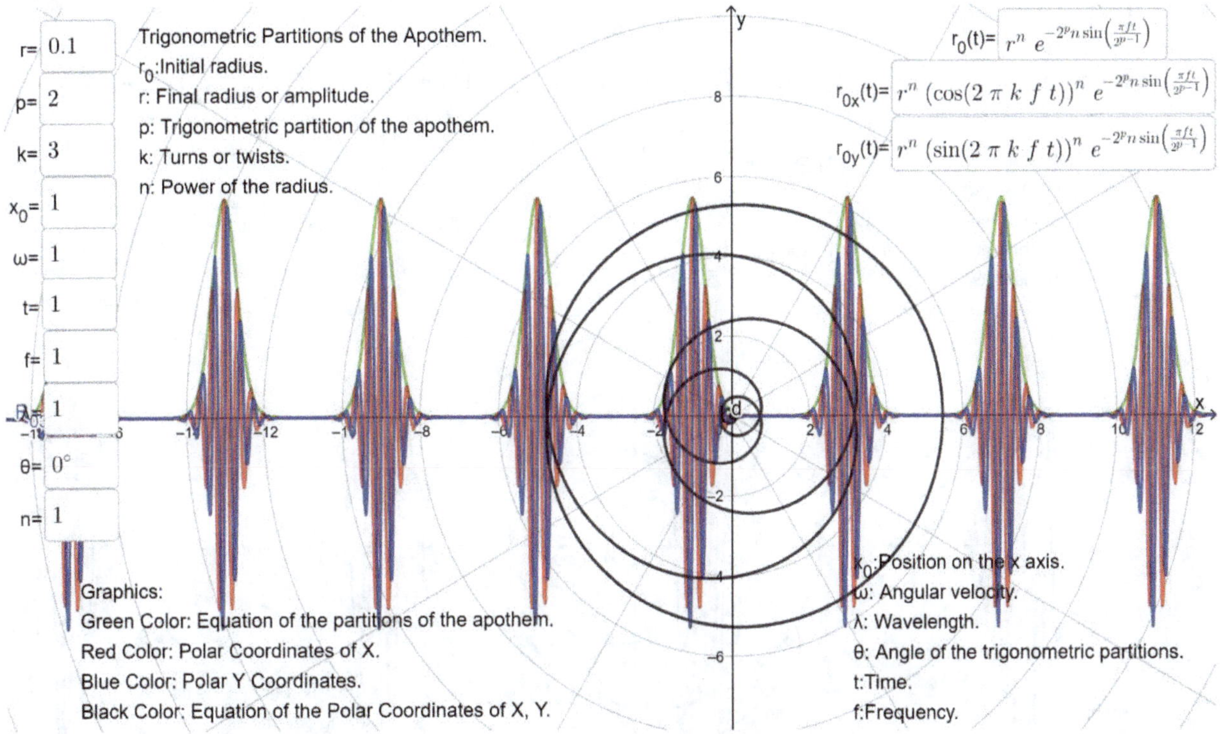

Note:This graph shows how the wavelets are generated by the powers n=1 with the equations of the initial radius as a function of time.

If we continue doing the same with the equation of the angle $\theta = k * x_0 - \omega * t$, if we substitute this equation in the function of the trigonometric partitions of the apothem in the equation of the radius as a function of the angle we will obtain:

$r_0(\theta) = r^n * e^{-2^p * n * \sin(\theta/2^p)}$.

$r_{0x}(\theta) = r^n * \cos^n(k * \theta) * e^{-2^p * n * \sin(\theta/2^p)}$.

$r_{0y}(\theta) = r^n * \sin^n(k * \theta) * e^{-2^p * n * \sin(\theta/2^p)}$.

These radius equations can be written as a function of the number of waves, the position on the x-axis, the angular velocity, and time; generating wavelets. If we want to move it to any position on the **y** axis, we only have to add a constant **g** to the equation of the trigonometric partitions of the radius.

Figure 2.15: General Equation Of Trigonometric Partitions Of The Apothem.

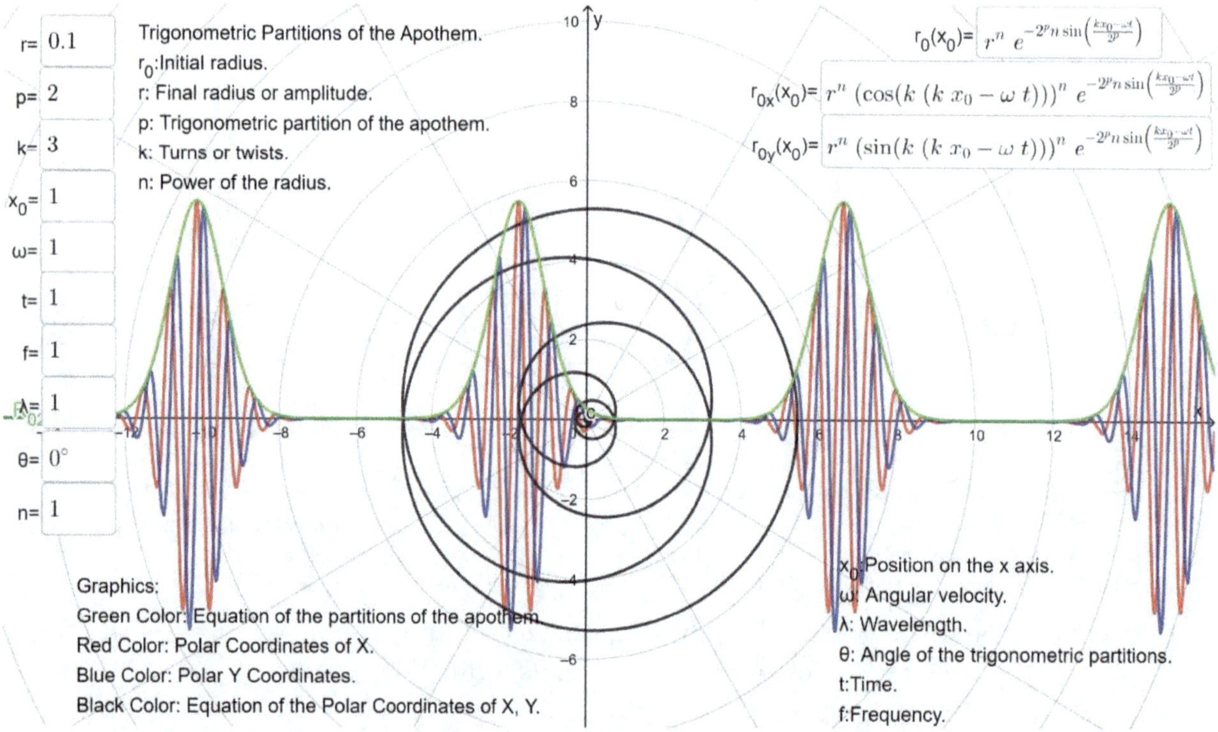

Note: This graph shows how the wavelets are generated by the powers n=1 with the equations of the initial radius as a function of the position on the **x** axis.

Finally, if we continue doing the same with the equation of the angle $\theta = 2 * \pi * x_0/\lambda - \omega * t$,with $k = 2 * \pi/\lambda$, if we substitute these equations in the function of the trigonometric partitions of the apothem in the equation of the radius as a function of the angle we will obtain:

$r_0(\theta) = r^n * e^{-2^P * n * \sin(\theta/2^P)}$.

$r_{0x}(\theta) = r^n * \cos^n(k * \theta) * e^{-2^P * n * \sin(\theta/2^P)}$.

$r_{0y}(\theta) = r^n * \sin^n(k * \theta) * e^{-2^P * n * \sin(\theta/2^P)}$.

These radius equations can be written as a function of wavelength, position on the x-axis, angular velocity, and time; generating wavelets. If we want to move it to any position on the **y** axis, we just have to add a constant **g** to the equation of the trigonometric partitions of the radius.

Figure 2.16: General Equation Of Trigonometric Partitions Of The Apothem.

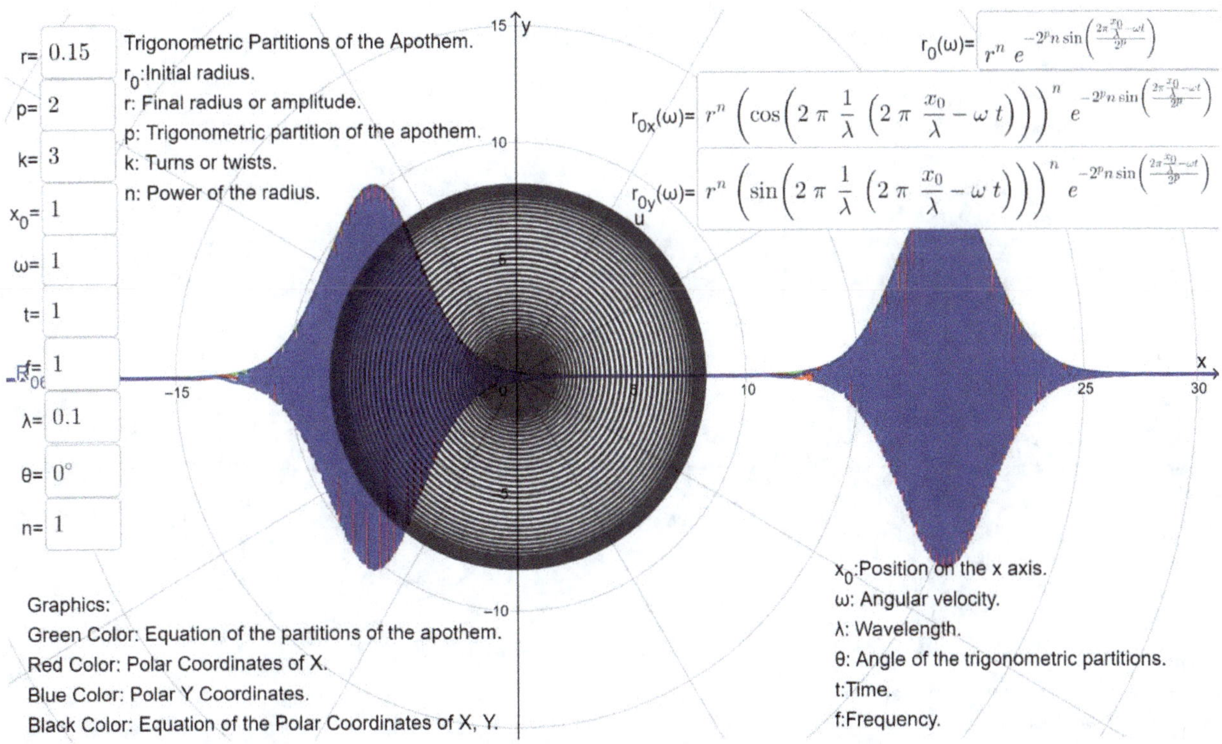

Note:This graph shows how the wavelets are generated by the powers n=1 with the equations of the initial radius as a function of the angular velocity ω.

2.6 Wavelets Transforms Of Trigonometric Partitions Of The Arrow.

If we substitute the equivalent equation of the angle as a function of angular velocity for time, in the general equation of the trigonometric partitions of the arrow we will obtain: The angle is in radians. $\theta = \omega * t$.

$r_0(\theta) = r^n * e^{2^P*n*\sin(\theta/2^P)-n*\theta}$.

$r_{0x}(\theta) = r^n * \cos^n(k*\theta) * e^{2^P*n*\sin(\theta/2^P)-n*\theta}$.

$r_{0y}(\theta) = r^n * \sin^n(k*\theta) * e^{2^P*n*\sin(\theta/2^P)-n*\theta}$.

These radius equations can be written as a function of angular velocity and time, generating wavelets. If we want to move it to any position on the **y** axis, we just have to add a constant **g** to the equation of the trigonometric partitions of the radius.

Figure 2.17: General Equation Of Trigonometric Partitions Of The Arrow.

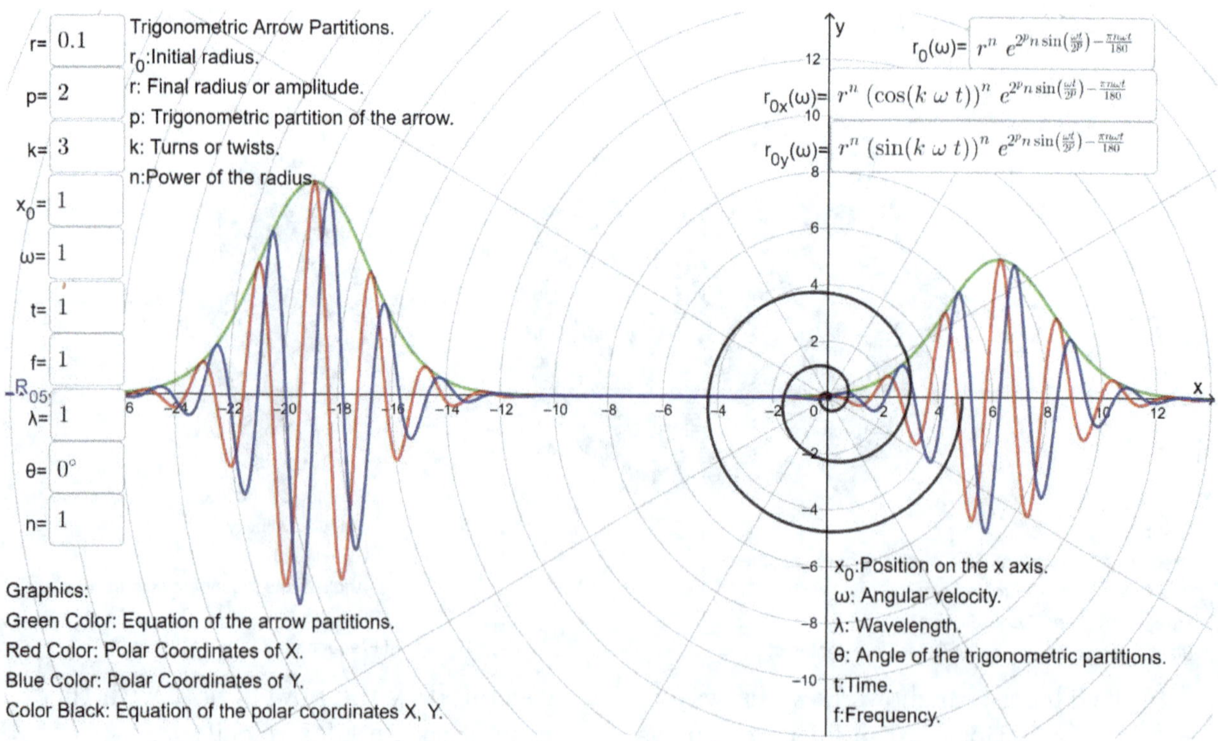

Note:This graph shows how the wavelets are generated by the powers n with the equations of the initial radius as a function of the angular velocity.

Also, we can do the same if we substitute; The angle in radians.

$\theta = 2 * \pi * f * t$ in the equations of the trigonometric partitions of the radius arrow as a function of the angle:

$r_0(\theta) = r^n * e^{2^P * n * \sin(\theta/2^P) - n*\theta}$.

$r_{0x}(\theta) = r^n * \cos^n(k * \theta) * e^{2^P * n * \sin(\theta/2^P) - n*\theta}$.

$r_{0y}(\theta) = r^n * \sin^n(k * \theta) * e^{2^P * n * \sin(\theta/2^P) - n*\theta}$.

These radius equations can be written as a function of frequency and time, generating wavelets. If we want to move it to any position on the **y** axis, we just have to add a constant **g** to the equation of the trigonometric partitions of the radius.

Figure 2.18: General Equation Of Trigonometric Partitions Of The Arrow.

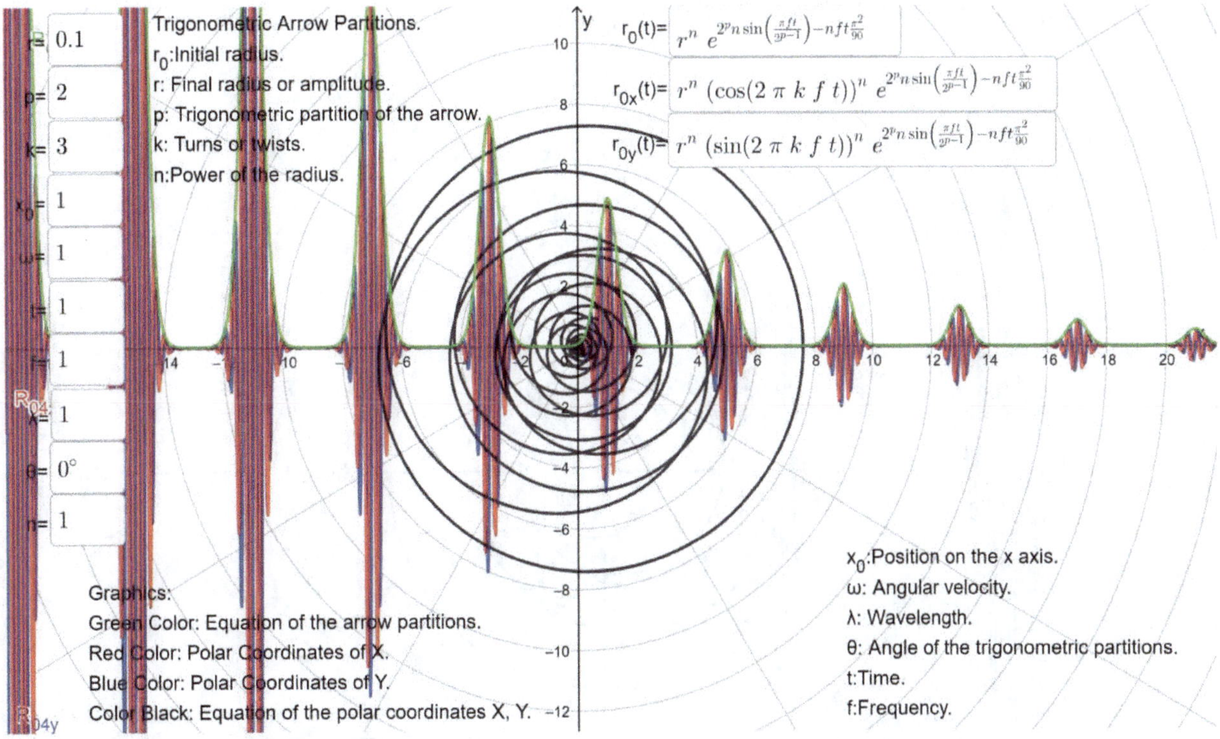

Note:This graph shows how the wavelets are generated by the powers n=1 with the equations of the initial radius as a function of time.

If we continue doing the same with the angle equation. The angle is in radians. $\theta = k*x_0 - \omega*t$, if we substitute this equation in the function of the trigonometric partitions of the arrow in the equation of the radius as a function of the angle we will obtain:

$r_0(\theta) = r^n * e^{2^P*n*\sin(\theta/2^P)-n*\theta}$.

$r_{0x}(\theta) = r^n * \cos^n(k*\theta) * e^{2^P*n*\sin(\theta/2^P)-n*\theta}$.

$r_{0y}(\theta) = r^n * \sin^n(k*\theta) * e^{2^P*n*\sin(\theta/2^P)-n*\theta}$.

These radius equations can be written as a function of the number of waves, the position on the x-axis, the angular velocity, and time; generating wavelets. If we want to move it to any position on the **y** axis, we only have to add a constant **g** to the equation of the trigonometric partitions of the radius.

Figure 2.19: General Equation Of Trigonometric Partitions Of The Arrow.

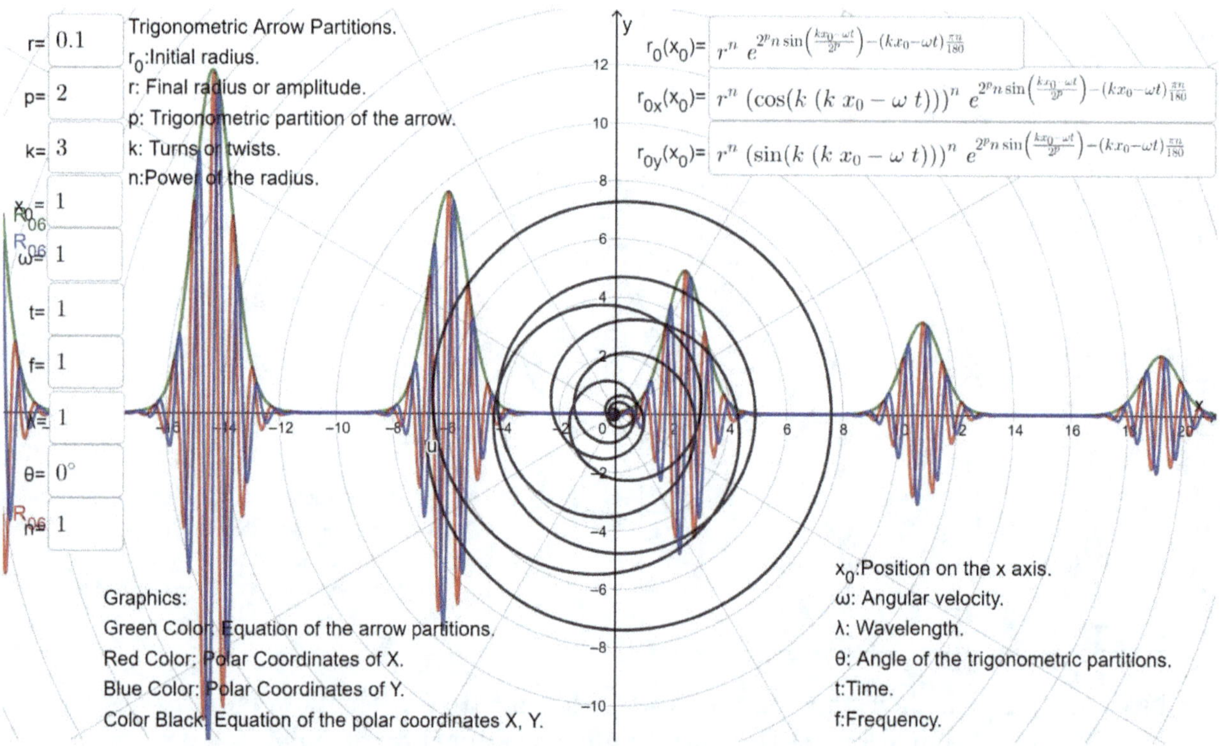

Note:This graph shows how the wavelets are generated by the powers n=1 with the equations of the initial radius as a function of the position on the **x** axis.

Finally, if we continue doing the same with the angle equation. The angle is in radians. $\theta = 2 * \pi * x_0/\lambda - \omega * t$,with $k = 2 * \pi/\lambda$, if we substitute these equations in the function of the trigonometric partitions of the arrow in the equation of the radius as a function of the angle we will obtain:

$r_0(\theta) = r^n * e^{2^P * n * \sin(\theta/2^P) - n*\theta}$.

$r_{0x}(\theta) = r^n * \cos^n(k * \theta) * e^{2^P * n * \sin(\theta/2^P) - n*\theta}$.

$r_{0y}(\theta) = r^n * \sin^n(k * \theta) * e^{2^P * n * \sin(\theta/2^P) - n*\theta}$.

These radius equations can be written as a function of wavelength, position on the x-axis, angular velocity, and time; generating wavelets. If we want to move it to any position on the **y** axis, we just have to add a constant **g** to the equation of the trigonometric partitions of the radius.

Figure 2.20: General Equation of Trigonometric Partitions of the Arrow.

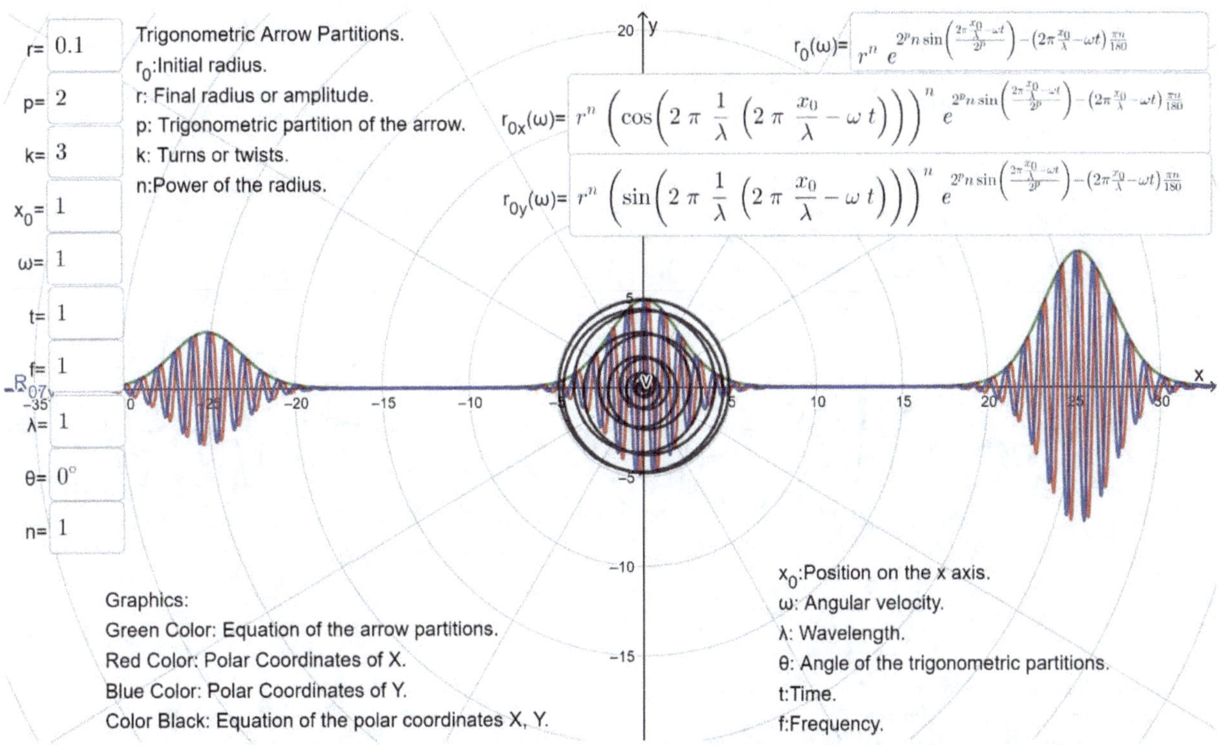

Note:This graph shows how the wavelets are generated by the powers n=1 with the equations of the initial radius as a function of the angular velocity ω.

2.7 Wavelets Transforms Of Trigonometric Partitions Of Radius Growth.

If we substitute the equivalent equation of the angle as a function of angular velocity for time, in the general equation of the trigonometric partitions of the growth of the radius we will obtain: The angle is in radians.$\theta = \omega * t$.

$r_0 = r^n * e^{2^{p-1}*n*\sin(\theta/2^{p-1})-n*\theta}$.

$r_{0x}(\theta) = r^n * \cos^n(k*\theta) * e^{2^{p-1}*n*\sin(\theta/2^{p-1})-n*\theta}$.

$r_{0y}(\theta) = r^n * \sin^n(k*\theta) * e^{2^{p-1}*n*\sin(\theta/2^{p-1})-n*\theta}$.

These radius equations can be written as a function of angular velocity and time, generating wavelets. If we want to move it to any position on the **y** axis, we just have to add a constant **g** to the equation of the trigonometric partitions of the radius.

Figure 2.21: General Equation Of Trigonometric Partitions Of Radius Growth.

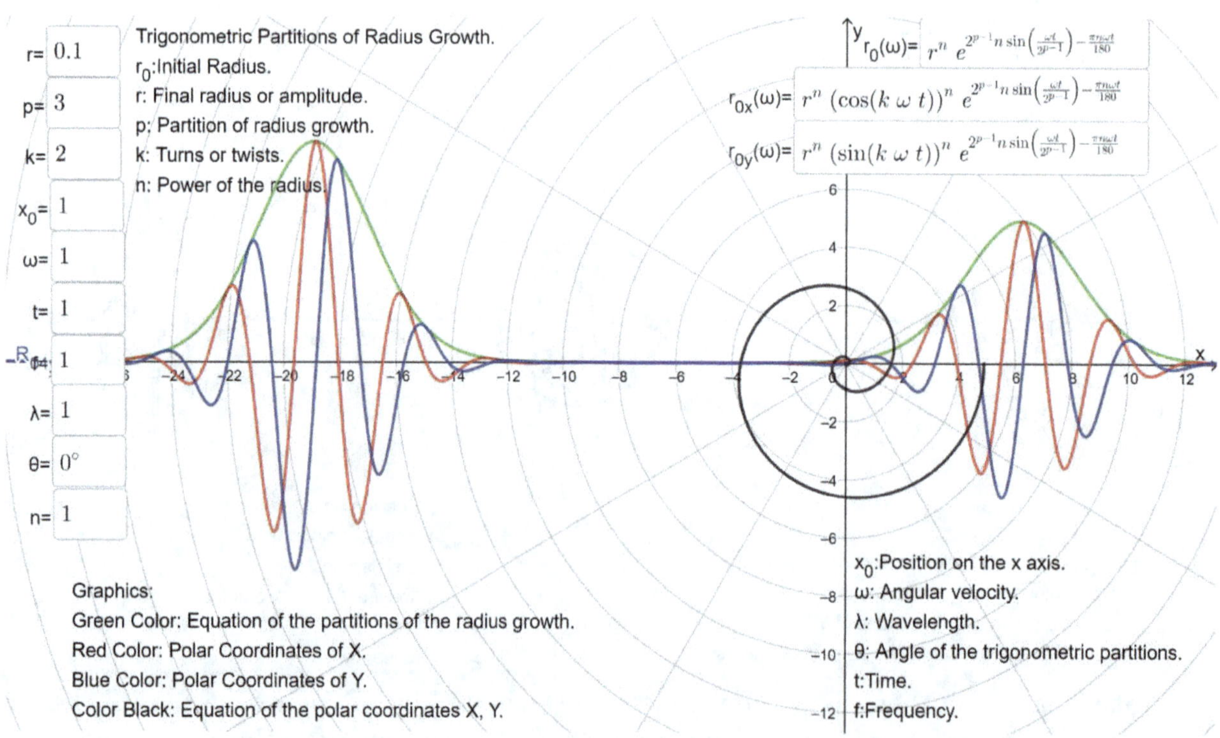

Note:This graph shows how the wavelets are generated by the powers n with the equations of the initial radius as a function of the angular velocity.

Also, we can do the same if we substitute; The angle in radians.

$\theta = 2 * \pi * f * t$ in the equations of the trigonometric partitions of the growth of x of the radius as a function of the angle:

$r_0(\theta) = r^n * e^{2^{p-1} * n * \sin(\theta/2^{p-1}) - n * \theta}$.

$r_{0x}(\theta) = r^n * \cos^n(k * \theta) * e^{2^{p-1} * n * \sin(\theta/2^{p-1}) - n * \theta}$.

$r_{0y}(\theta) = r^n * \sin^n(k * \theta) * e^{2^{p-1} * n * \sin(\theta/2^{p-1}) - n * \theta}$.

These radius equations can be written as a function of frequency and time, generating wavelets. If we want to move it to any position on the **y** axis, we just have to add a constant **g** to the equation of the trigonometric partitions of the radius.

Figure 2.22: General Equation Of Trigonometric Partitions Of Radius Growth.

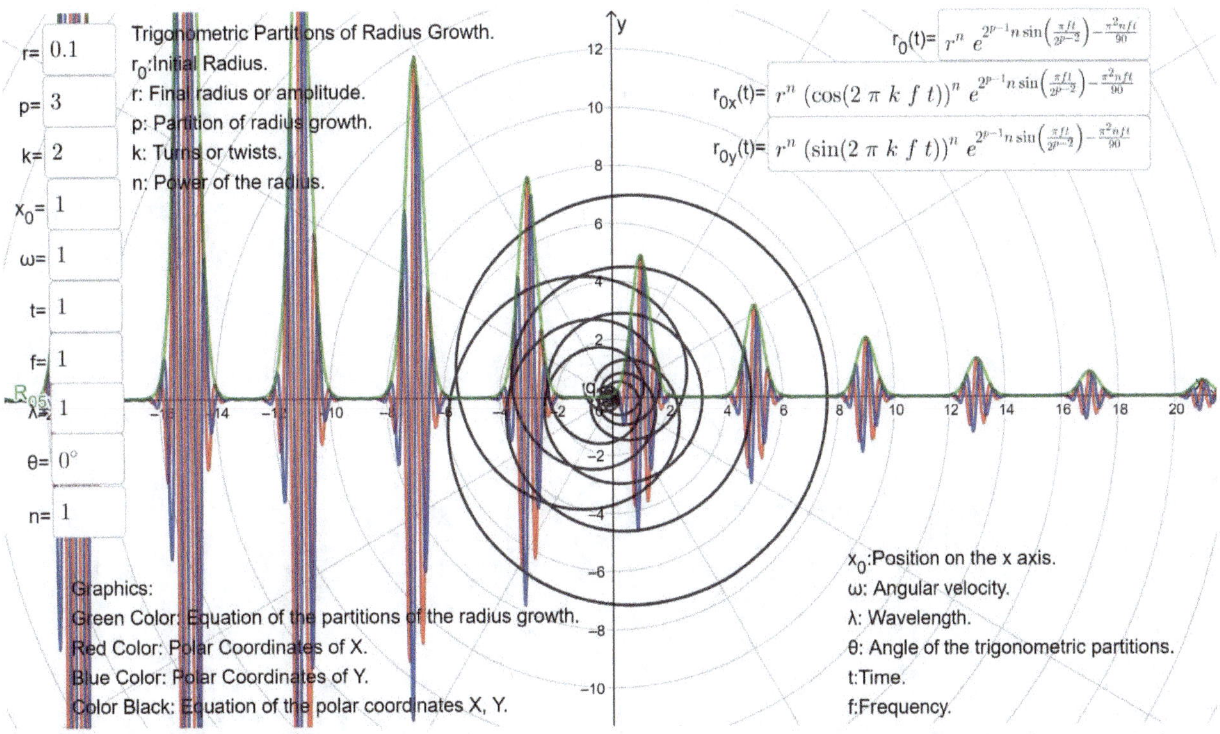

Note:This graph shows how the wavelets are generated by the powers n=1 with the equations of the initial radius as a function of time.

If we continue doing the same with the angle equation. The angle is in radians. $\theta = k * x_0 - \omega * t$, if we substitute this equation in the function of the trigonometric partitions of the growth of the radius in the equation of the radius as a function of the angle we will obtain:

$r_0(\theta) = r^n * e^{2^{p-1} * n * \sin(\theta/2^{p-1}) - n*\theta}$.

$r_{0x}(\theta) = r^n * \cos^n(k * \theta) * e^{2^{p-1} * n * \sin(\theta/2^{p-1}) - n*\theta}$.

$r_{0y}(\theta) = r^n * \sin^n(k * \theta) * e^{2^{p-1} * n * \sin(\theta/2^{p-1}) - n*\theta}$.

These radius equations can be written as a function of the number of waves, the position on the x-axis, the angular velocity, and time; generating wavelets. If we want to move it to any position on the **y** axis, we only have to add a constant **g** to the equation of the trigonometric partitions of the radius.

Figure 2.23: General Equation Of Trigonometric Partitions Of Radius Growth.

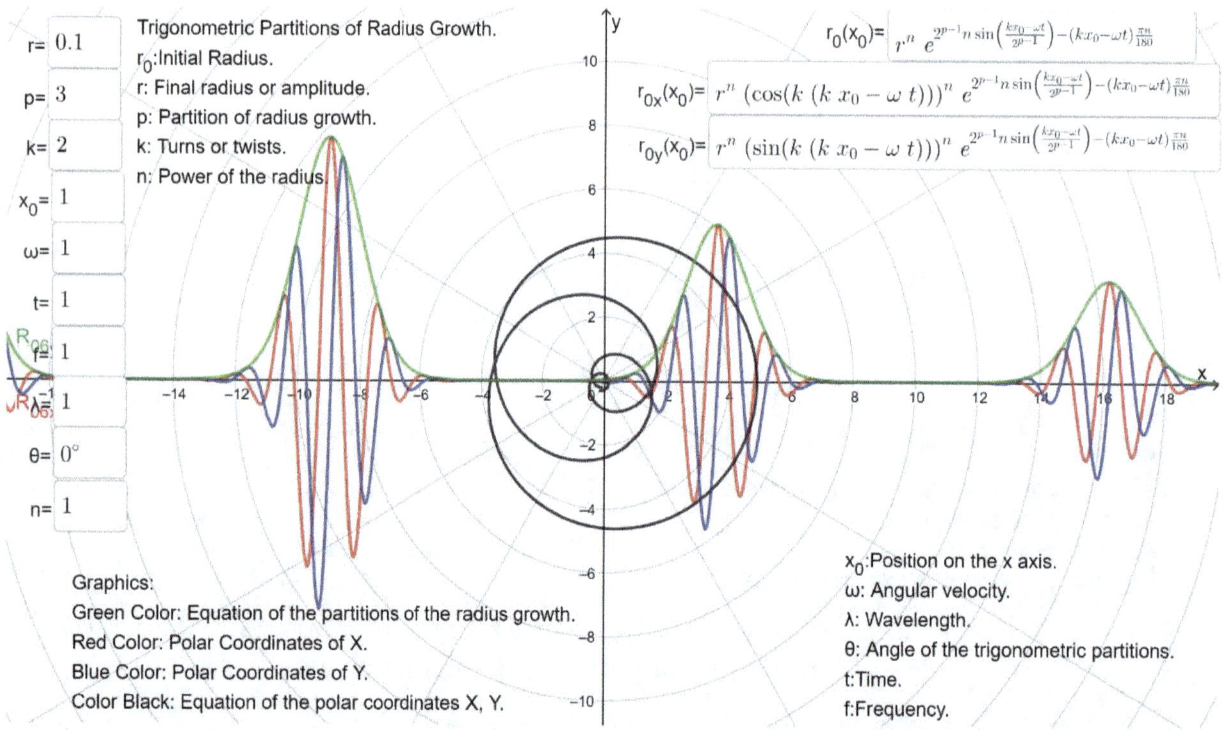

Note:This graph shows how the wavelets are generated by the powers n=1 with the equations of the initial radius as a function of the position on the **x** axis.

Finally, if we continue doing the same with the angle equation. The angle is in radians. $\theta = 2 * \pi * x_0/\lambda - \omega * t$, with $k = 2 * \pi/\lambda$, if we substitute these equations in the function of the trigonometric partitions of the growth of the radius in the equation of the radius as a function of the angle we will obtain:

$r_0(\theta) = r^n * e^{2^{p-1} * n * \sin(\theta/2^{p-1}) - n * \theta}$.

$r_{0x}(\theta) = r^n * \cos^n(k * \theta) * e^{2^{p-1} * n * \sin(\theta/2^{p-1}) - n * \theta}$.

$r_{0y}(\theta) = r^n * \sin^n(k * \theta) * e^{2^{p-1} * n * \sin(\theta/2^{p-1}) - n * \theta}$.

These radius equations can be written as a function of wavelength, position on the x-axis, angular velocity, and time; generating wavelets. If we want to move it to any position on the **y** axis, we just have to add a constant **g** to the equation of the trigonometric partitions of the radius.

Figure 2.24: General Equation Of Trigonometric Partitions Of Radius Growth.

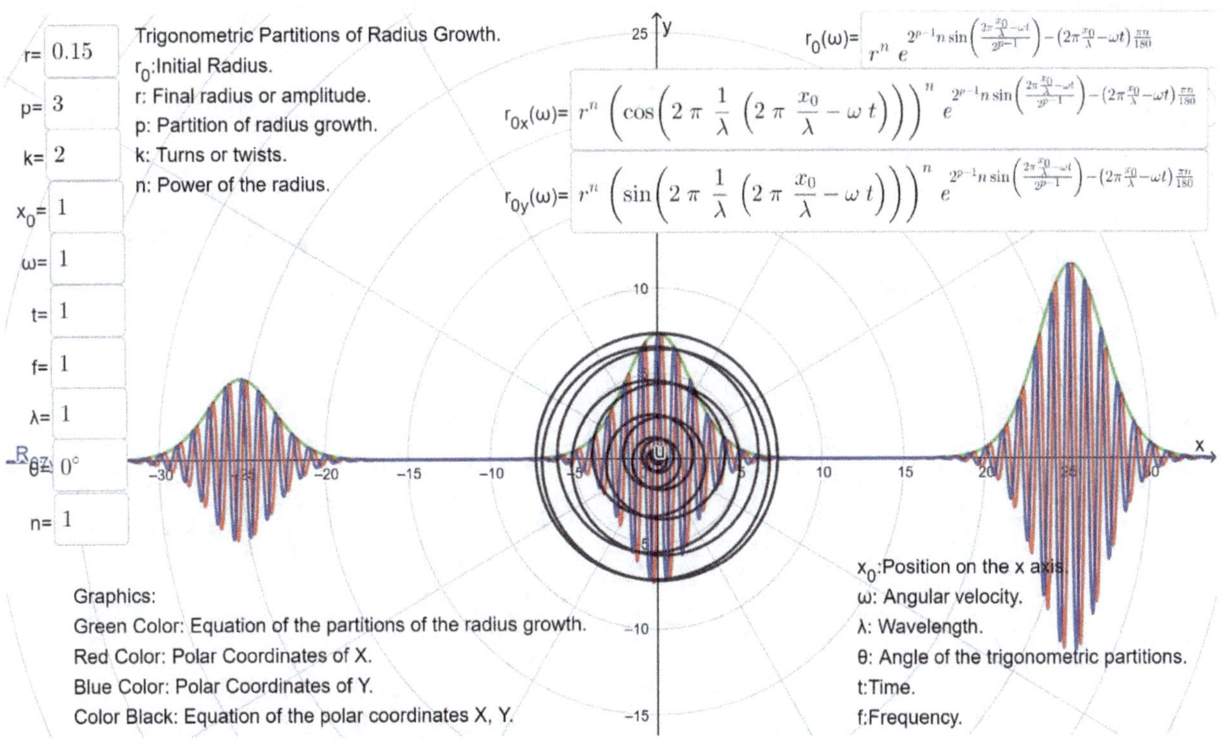

Note:This graph shows how the wavelets are generated by the powers n=1 with the equations of the initial radius as a function of the angular velocity ω.

2.8 Application Of Wavelets Transforms Of Trigonometric Partitions To Circular Motion.

Applying the Theory Of Trigonometric Partitions to a particle, that is moving in a circular path, moving along its perimeter or circular sector, with an angular displacement, developing a tangential velocity as a function of time; which is perpendicular to the radius, we can use the trigonometric partitions of the Circular Sector $r_0 = r * e^{-2^{2*p+1}*\sin^2(\theta/2^{p+1})}$ with its initial radius and apply it to the classical physics concept of uniform circular motion. Classical physics teaches us; the definition of tangential velocity as equivalent to angular velocity times radius. Where the compression radius is the equation of the trigonometric partitions of the circular sector; This tells us that if the particle loses speed until it reaches zero, it can fall in a spiral towards the center of its generated circumference. $v_t = \omega * r_0$. (Serway et al., 2018)

Figure 2.25: Transforms Of The Trigonometric Partitions Of The Circular Sector.

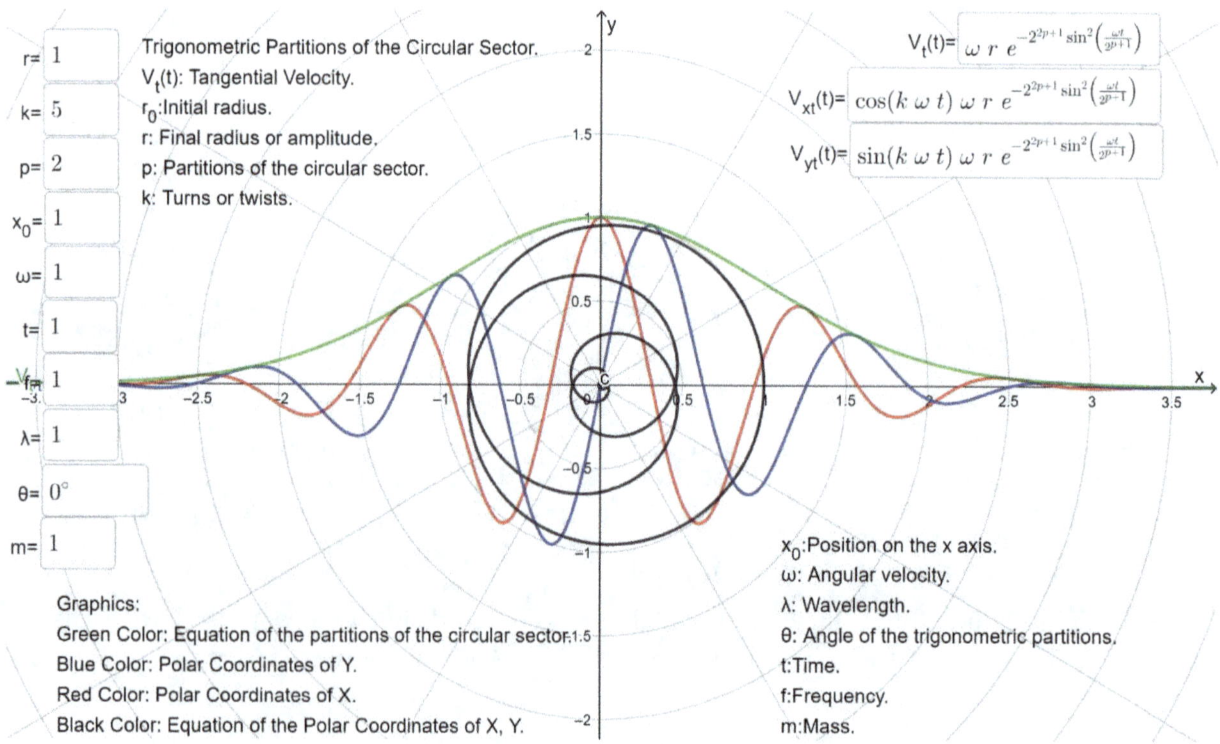

Note: This graph shows how the wavelets are generated by the applications of trigonometric partitions to the concepts of classical physics of tangential velocity.

The normal or centripetal acceleration as a function of time; which is negative, indicating that it is towards the center of the circle, due to the square of the angular velocity multiple of the radius, it generates the square power of the tangential velocity proportional to the inverse of the radius, resulting in a centripetal force exerted in the movement circular particle: $a_c = \omega^2 * r_0 = v_t^2/r_0$. Substituting the equation of the trigonometric partitions of the circular sector $r_0 = r * e^{-2^{2*p+1}*\sin^2(\theta/2^{p+1})}$, in the centripetal acceleration equation as a function of time, which can also be written as a function of angular velocity and obtain wavelets, the following equations will be obtained with their graphs as shown in the following figure 2.26:

Figure 2.26: Transforms Of The Trigonometric Partitions Of The Circular Sector.

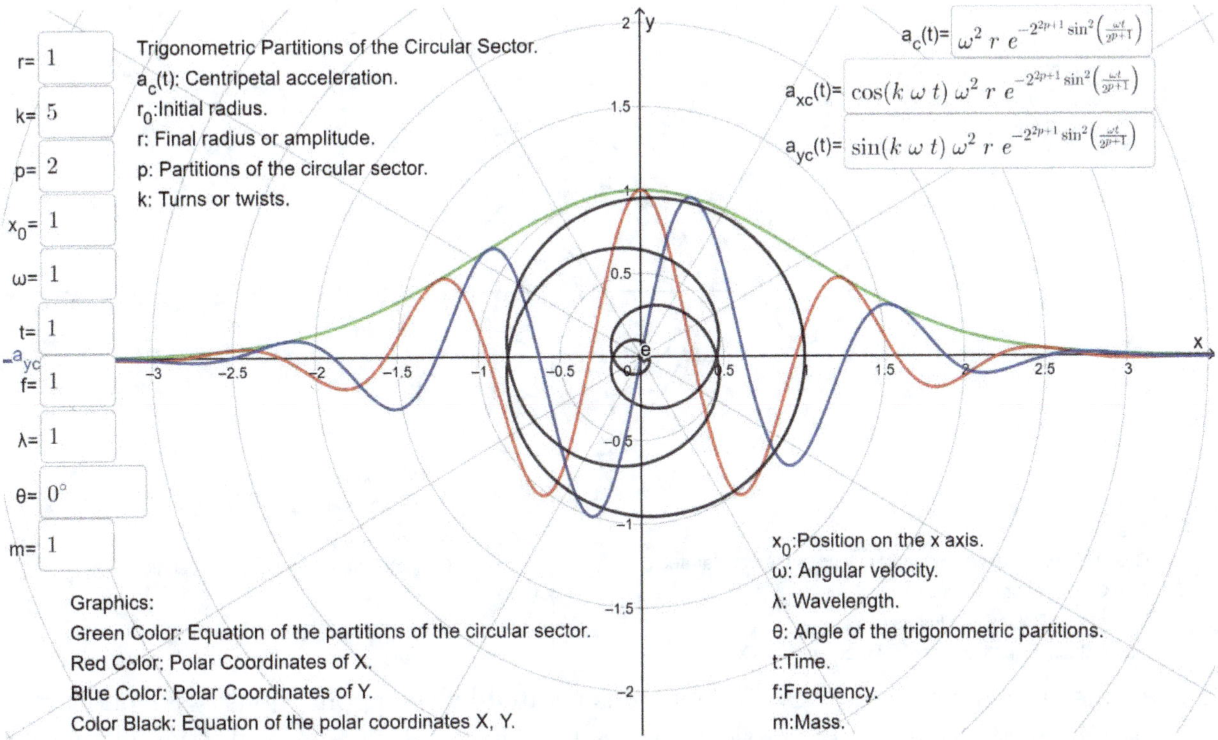

Note: This graph shows how the wavelets are generated by the applications of trigonometric partitions to the concepts of classical physics of centripetal acceleration as a function of time.

The centripetal force exerted on the particle as a function of time is the product of the mass times the centripetal acceleration of the particle that is performing a uniform circular movement and it is this force that attracts it towards the center of its circular movement, in the same direction of the radio. $F_c = m * a_c$. Substituting the equation of the trigonometric partitions of the circular sector $r_0 = r * e^{-2^{2*p+1} * \sin^2(\theta/2^{p+1})}$, in the equation of centripetal force as a function of time, which can also be written as a function of angular velocity and obtain wavelets, the following equations will be obtained with their graphs as shown in the following figure 2.27:

Figure 2.27: Transforms Of The Trigonometric Partitions Of The Circular Sector.

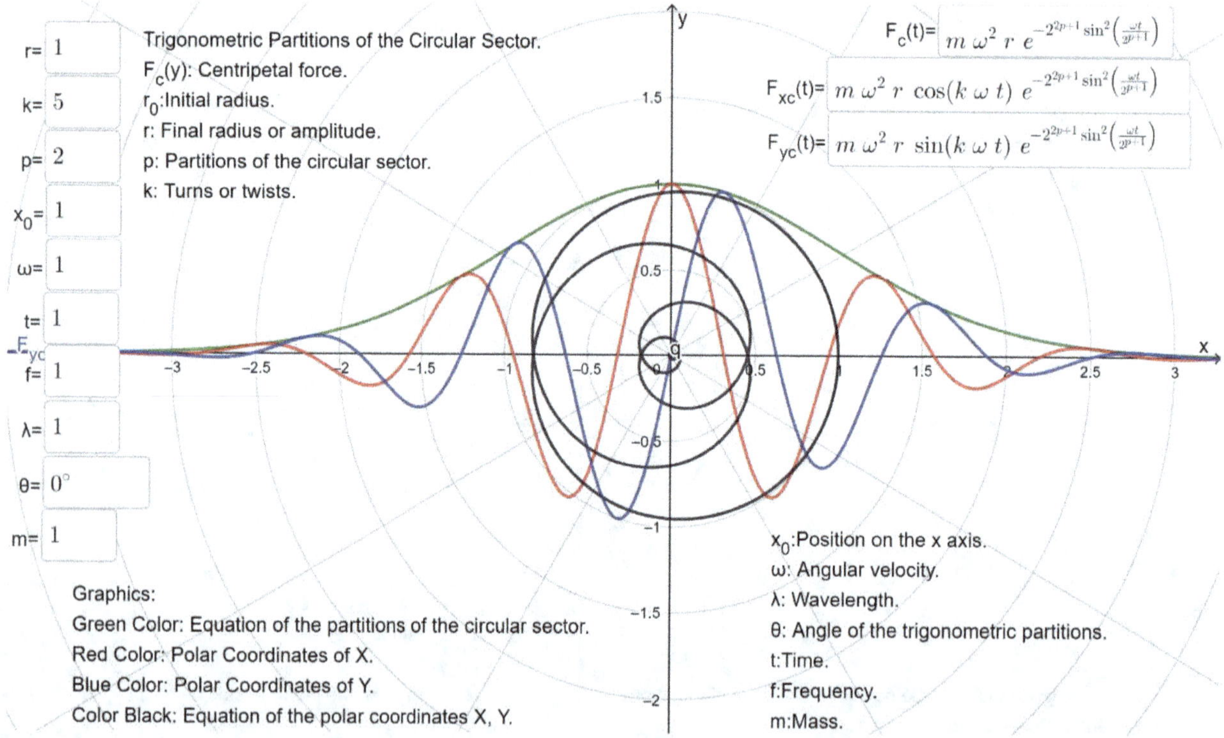

Note:This graph shows how the wavelets are generated by the applications of trigonometric partitions to the concepts of classical physics to centripetal force as a function of time.

Torque is the multiple of the centripetal force, exerted in the same direction as the radius, to generate the uniform circular motion of the particle. $\tau = F_c * r_0$. Substituting the equation of the trigonometric partitions of the circular sector $r_0 = r * e^{-2^{2*p+1}*\sin^2(\theta/2^{p+1})}$, in the torque equation as a function of time, which can also be written as a function of angular velocity and obtain wavelets, the following equations will be obtained with their graphs as shown in the following figure 2.28:

Figure 2.28: Transforms Of The Trigonometric Partitions Of The Circular Sector.

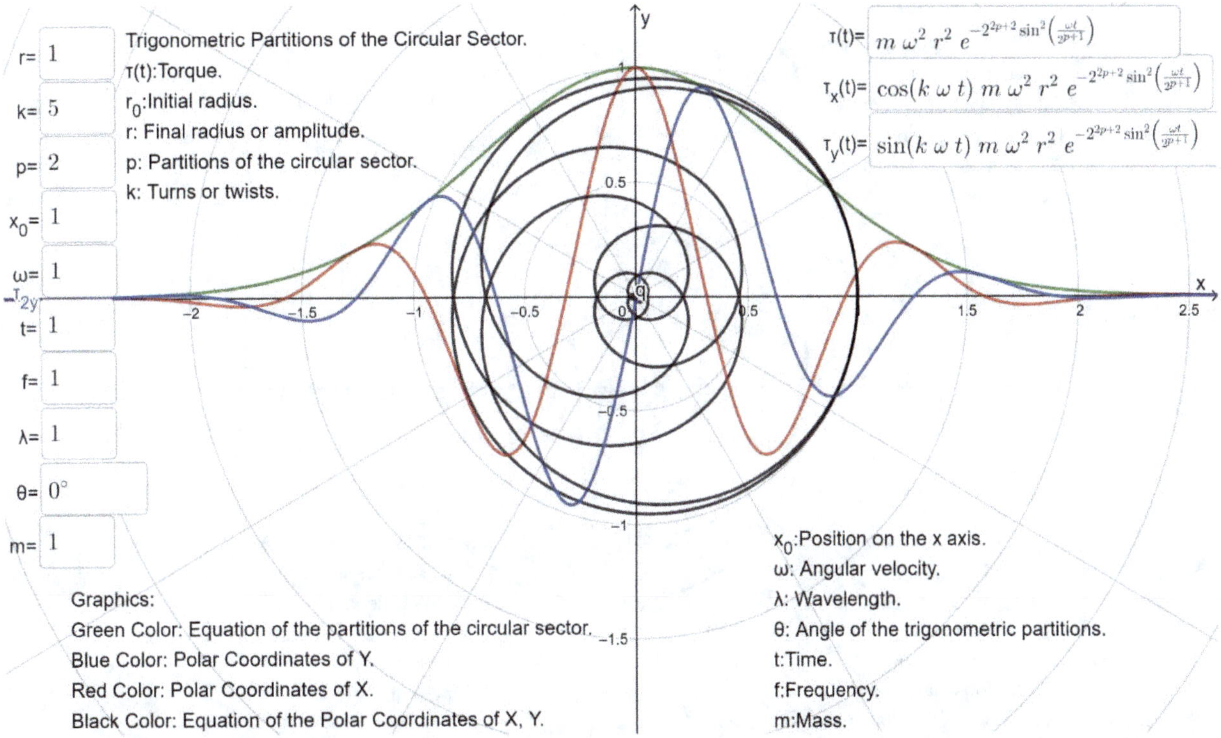

Note:This graph shows how the wavelets are generated by the applications of trigonometric partitions to the concepts of classical physics of torque as a function of time.

Linear momentum is the multiple of the tangential velocity over the mass, which generates the momentum to generate the initial circular motion of the particle. $\rho = m * v_t$. Substituting the equation of the trigonometric partitions of the circular sector $r_0 = r * e^{-2^{2*p+1}*\sin^2(\theta/2^{p+1})}$, In the equation of linear momentum as a function of time, which can also be written as a function of angular velocity and obtain wavelets, the following equations will be obtained with their graphs as shown in the following figure 2.29:

Figure 2.29: Transforms Of The Trigonometric Partitions Of The Circular Sector.

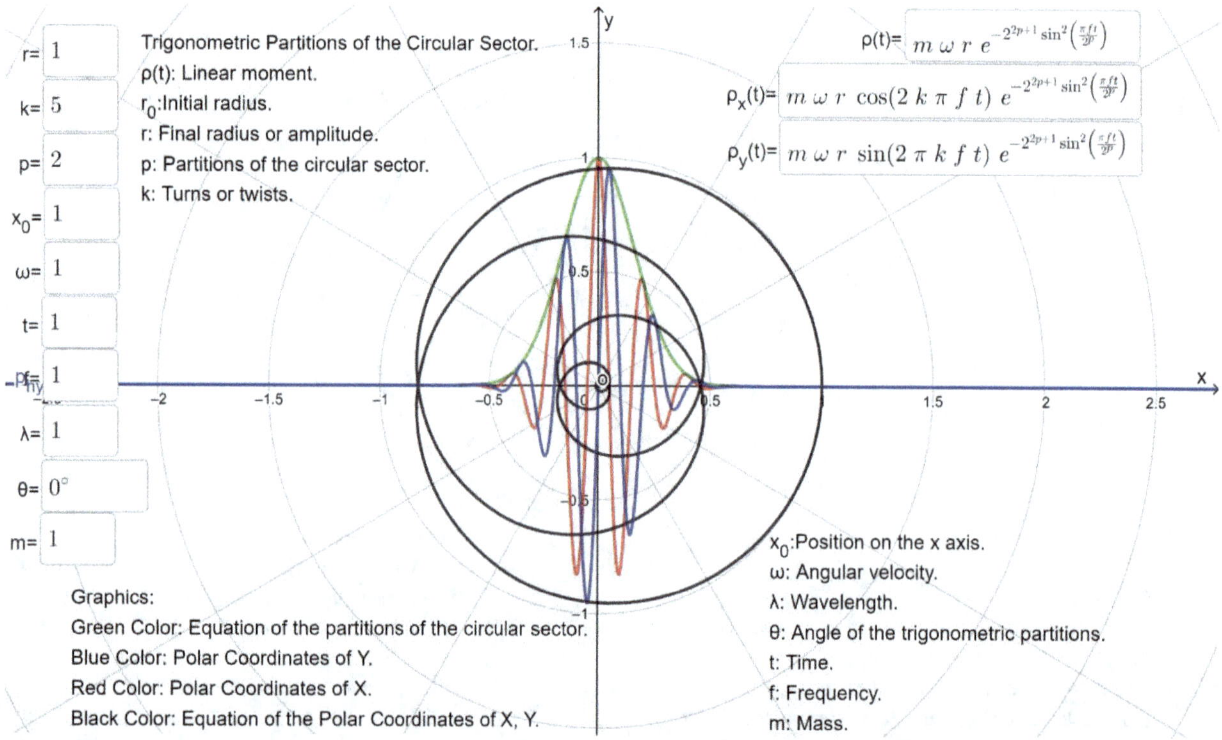

Note:This graph shows how the wavelets are generated by the applications of trigonometric partitions to the concepts of classical physics of linear momentum as a function of time.

Inertial moment is the concentration of mass, perpendicular to the square power of the radius, and is what causes its physical moment to begin, causing the rotation of the particle. $I_0 = m * r_0^2$. Substituting the equation of the trigonometric partitions of the circular sector $r_0 = r * e^{-2^{2*p+1} * \sin^2(\theta/2^{p+1})}$, in the equation of the inertial moment as a function of time, which can also be written as a function of the angular velocity and obtain wavelets, the following equations will be obtained with their graphs as shown in the following figure 2.30:

Figure 2.30: Transforms Of The Trigonometric Partitions Of The Circular Sector.

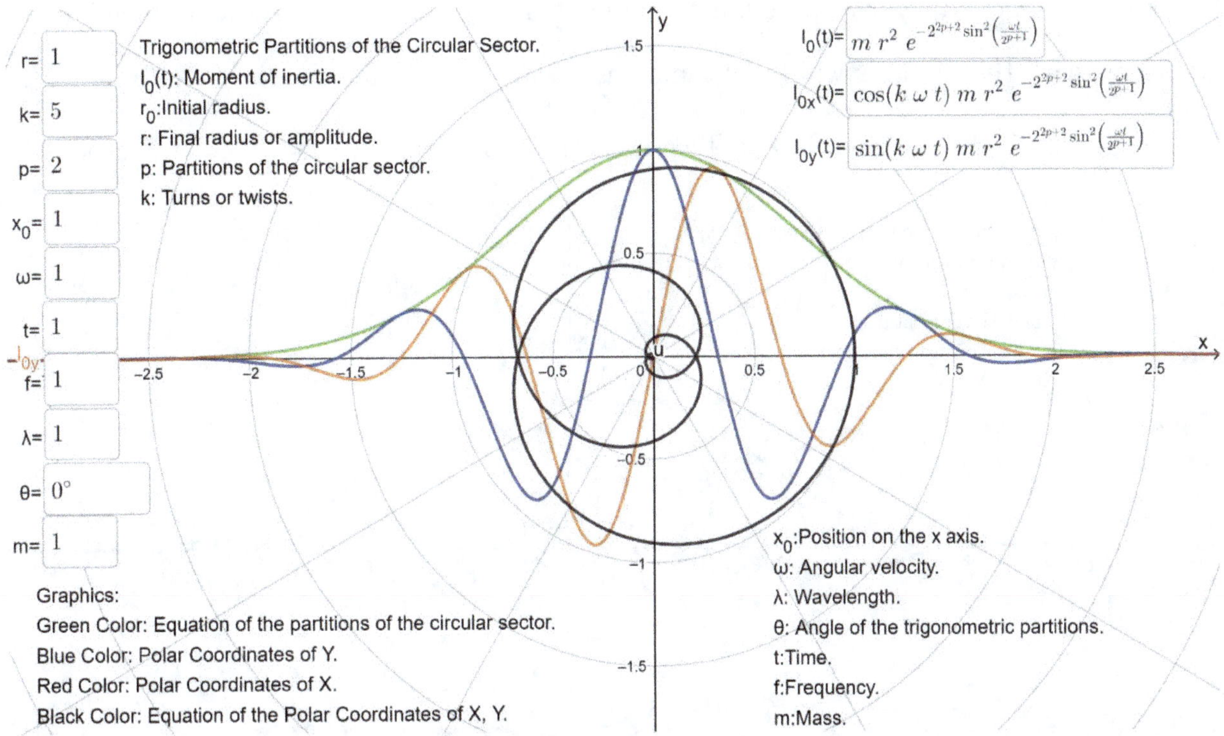

Note:This graph shows how the wavelets are generated by the applications of trigonometric partitions to the concepts of classical physics of inertial moment as a function of time.

This same process that has been applied to uniform circular motion using the circular sector equation, in relation to the trigonometric partitions, can be applied to all the trigonometric partition equations.

Angular momentum is the angular velocity and mass perpendicular to the square power of the radius, which produces the uniform circular motion of the particle around its axis of rotation. The angular momentum perpendicular to the circumference is related to its inertial moment, a product of the angular velocity of the particle, to generate uniform circular motion in its central axis of rotation. $L = m * r_0^2 * \omega = I_0 * \omega$.
Substituting the equation of the trigonometric partitions of the circular sector $r_0 = r * e^{-2^{2*p+1} * \sin^2(\theta/2^{p+1})}$, In the equation of angular momentum as a function of time, which can also be written as a function of angular velocity and obtain wavelets, the following equations with their graphs will be obtained as shown in the following figure 2.31:

Figure 2.31: Transforms Of The Trigonometric Partitions Of The Circular Sector.

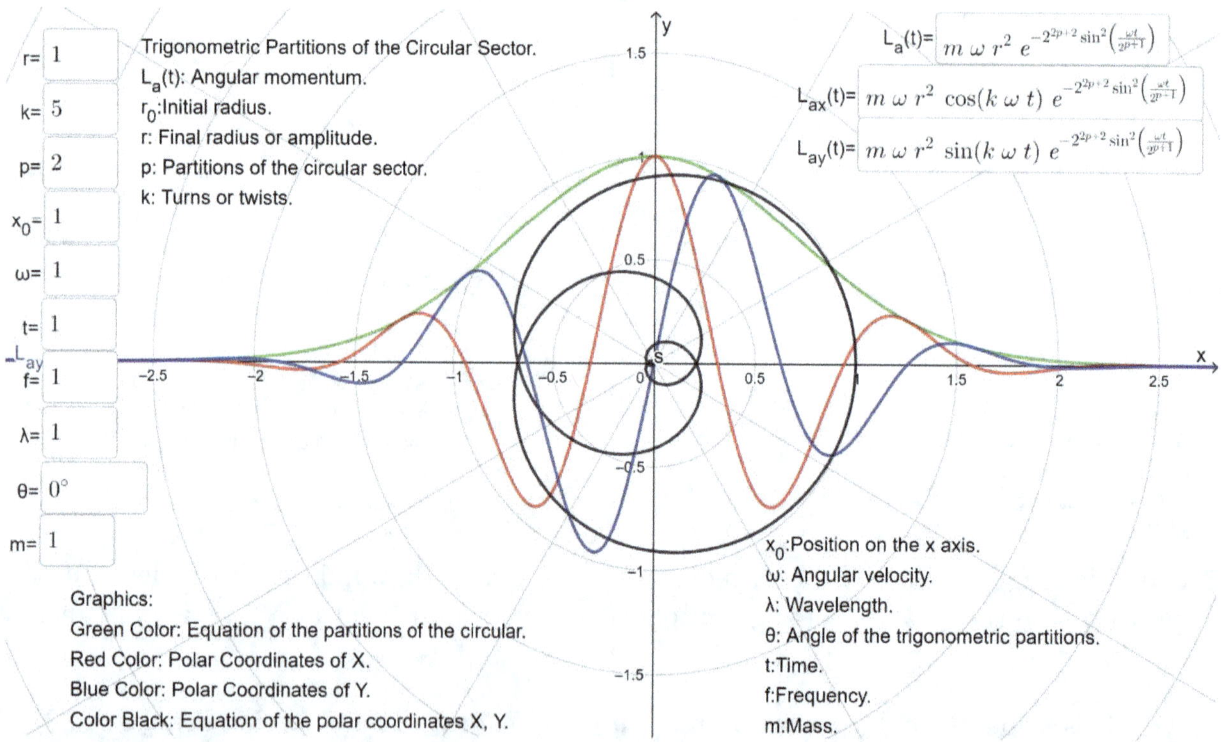

Note: This graph shows how the wavelets are generated by the applications of trigonometric partitions to the concepts of classical physics of angular momentum as a function of time.

2.9 Trigonometric Partitions Identities Of Mid-Angles:

The unit circle is continuously divided into several halves, yielding new trigonometric identities for mean angles. With this methodology of continuously dividing the circle, we can develop a new theory to analyze the Cartesian plane with trigonometric partitions. With this theory of trigonometric partitions, greater precision can be given to the equation, yielding exact results. Similarly, it is possible to study the pattern of a mathematical function, search for the solution of a difficult mathematical equation using trial and error methodology, and finally, develop a simpler and more understandable mathematical equation for the wavelet transform. If we continuously make radial or diametral cuts on a circle, we can divide it into successive halves. This physical effect of continuously cutting the circle is called "Trigonometric Partitions." This physical effect of continuous cuts can be executed radially for the apothem, arrow, and radius growth. Another way to make cuts is in the direction of the chord, circular sector, and along the X and Y axes. The cut of the chord is called "Chord Trigonometric Partition," the cut of the circular sector is called "Circular Sector Trigonometric Partition," the cut of the arrow is called "Arrow Trigonometric Partition," the cut of the apothem is called "Apothem Trigonometric Partition," and so on. This physical effect of trigonometric cuts will be denoted by the letter "p." The physical effect of "p" can be considered positive or negative depending on the direction of the cut or rotation of its angular velocity. If we make radial cuts of the circle, when "p" equals zero, then its physical form is a circle; if "p" equals one, then the circle has been divided into two halves; if "p" equals two, the circle has been divided into four equal parts, and so on. This is the general trigonometric identity for trigonometric partitions (p). Where p represents the physical effect of dividing or cutting a circle into infinite parts.

With $p = 0$, $2 * \sin(\theta/2) = \sqrt{2} * \sqrt{1 - \cos(\theta)}$ and $2 * \sin(\theta/2^{p+1}) = \sqrt{2} * \sqrt{1 - \cos(\theta/2^p)}$.

Where $2 * \sin^2(\theta/2^{p+1}) = 1 - \cos(\theta/2^p)$, Therefore,$\cos(\theta/2^p) = 1 - 2 * \sin^2(\theta/2^{p+1})$.

Defining p as the trigonometric partitions in a circle; p can be any number. The following general trigonometric identity for mid-angles can be demonstrated with the following graph, starting with p=0 then; $2 * \sin(\theta/2^{p+1}) = \sqrt{2} * \sqrt{1 - \cos(\theta/2^p)}$, solving for the cosine of the angle we can obtain: $\cos(\theta/2^p) = 1-2*\sin^2(\theta/2^{p+1})$ or $\sin(\theta/2^{p+1}) = \sqrt{2} * \sqrt{1 - \cos(\theta/2^p)}/2$. Similarly, this concept applies to the trigonometric identity of the unit circle $\cos^2(\theta/2^{p+1}) + sin^2(\theta/2^{p+1}) = 1$. If we substitute $\sin^2(\theta/2^{p+1}) = [1 - \cos(\theta/2^p)]/2$ into the general trigonometric identity for the unit circle, we will get $\cos(\theta/2^p) = 2*\cos^2(\theta/2^{p+1}) - 1$. For example, as a reference point for the trigonometric partitions of the circular sector, the chord, and the radius growth, we take the initial angle Theta $\theta_0 = 0$. Also, the final angle Theta can be considered as $\theta = 2 * \pi$, in case more than 360° degrees or several turns are desired, the letter n will be used to define the equation of the angle Theta as $\theta = 2 * \pi * n$. These trigonometric partition equations of the angle can be used about the number of turns, angular velocity, frequency, time, position x, wavelength, and number of waves $\theta = 2 * \pi * n = \omega * t = 2 * \pi * f * t = k * x - \omega * t$ with $k = 2 * \pi/\lambda$. Rewriting the formulas, depending on the study being conducted, for example; using with its number of turns n: If we use $\theta = 2 * \pi * n$, in the following equation $\cos(\theta/2^p) = 1 - 2 * \sin^2(\theta/2^{p+1})$ we will obtain $cos(\pi * n/2^{p-1}) = 1 - 2 * \sin^2(\pi * n/2^p)$, in $\cos(\theta/2^p) = 2 * \cos^2(\theta/2^{p+1}) - 1$ we will obtain $\cos(\pi * n/2^{p-1}) = 2 * \cos^2(\pi * n/2^{p-1})$. Therefore, we can do the same with the angular velocity and time $\theta = \omega * t$ substituting in $\cos(\theta/2^p) = 1 - 2 * \sin^2(\theta/2^{p+1})$ we will obtain $\cos(\omega * t/2^p) = 1 - 2 * \sin^2(\omega * t/2^{p+1})$ or $\cos(\omega * t/2^p) = 2 * \cos^2(\omega * t/2^{p+1}) - 1$.

Circular diagram to develop general mid-angle trigonometric identities.

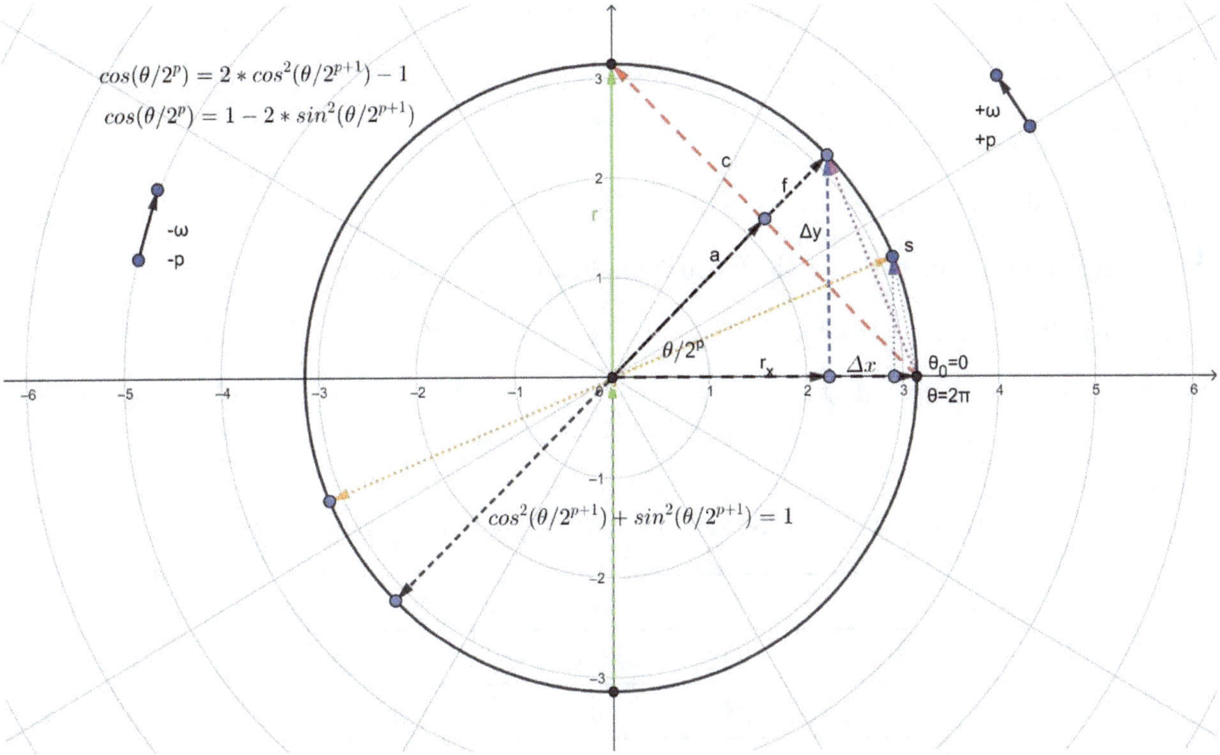

Figure 2.32: With this graph, all equations for mid-angles in a general form can be developed and the type of trigonometric cuts that can be made can be understood.

This advanced mathematical concept can be applied in classical and quantum physics to analyze all equations containing a radius in their formulas. Similarly, we can do the same with frequency and time $\theta = 2 * \pi * f * t$ by substituting into $\cos(\theta/2^p) = 1 - 2 * \sin^2(\theta/2^{p+1})$ to obtain $\cos(\pi * f * t/2^{p-1}) = 1 - 2 * \sin^2(\pi * f * t/2^p)$ or $\cos(\pi * f * t/2^{p-1}) = 2 * \cos^2(\pi * f * t/2^p) - 1$. Finally, we can do this with the number and wavelength in position x as a function of time $\theta = k * x - \omega * t$ with $k = 2 * \pi/\lambda$ by substituting into $\cos(\theta/2^p) = 1 - 2 * \sin^2(\theta/2^{p+1})$ to obtain $\cos[(k * x - \omega * t)/2^p] = 1 - 2 * \sin^2[(k * x - \omega * t)/2^{p+1}]$ or $\cos[(k * x - \omega * t)/2^p] = 2 * \cos^2[(k * x - \omega * t)/2^{p+1}] - 1$. Lastly, we have three general trigonometric identities for mid-angles in terms of trigonometric partitions, which are:

$2 * \cos(\theta/2^{p+1}) = \sqrt{2} * \sqrt{[1 + \cos(\theta/2^p)]}$,

$2 * \sin(\theta/2^{p+1}) = \sqrt{2} * \sqrt{[1 - \cos(\theta/2^p)]}$,

$\cos^2(\theta/2^{p+1}) + \sin^2(\theta/2^{p+1}) = 1,$

Regarding these equations, if we change the base (b) 2 to base (b) 3, this definition still holds for double trigonometric partitions, but not for triple trigonometric partitions:

$\cos^2(\theta/3^{p+1}) + \sin^2(\theta/3^{p+1}) = 1,$

$\cos^2(\theta/b^p) + \sin^2(\theta/b^p) = 1,$

$\cos^2(\theta/2^p) + \sin^2(\theta/2^p) = 1,$

If the trigonometric identities hold for any angle value and partitions.

$2 * \cos^2(\theta/b^{p+1}) = 1 + \cos(\theta/b^p).$

$2 * \sin^2(\theta/b^{p+1}) = 1 - \cos(\theta/b^p).$

Table 2.1: Proving Trigonometric Identities With Values.

$\theta=$	35	
b=	2	
p=	0	Sums
$\cos^2(\theta/b^p)$	0.671010072	1.000000000
$\sin^2(\theta/b^p)$	0.328989928	
$\cos^2(\theta/b^{p+1})$	0.909576022	1.000000000
$\sin^2(\theta/b^{p+1})$	0.090423978	
$2 * \cos^2(\theta/b^{p+1})$	1.819152044	
$1 + \cos(\theta/b^p)$	*1.819152044*	
$2 * \sin^2(\theta/b^{p+1})$	0.180847956	
$1 - \cos(\theta/b^p)$	*0.180847956*	

$$\cos^2(\theta/b^{p+1}) + \sin^2(\theta/b^{p+1}) = 1,$$

$\cos^2(\theta/3^{p+1}) + \sin^2(\theta/3^{p+1}) = 1,$

The trigonometric identities are not fulfilled.

$2 * \cos^2(\theta/b^{p+1}) \neq 1 + \cos(\theta/b^p).$

$2 * \sin^2(\theta/b^{p+1}) \neq 1 - \cos(\theta/b^p).$

Table 2.2: Proving Trigonometric Identities With Values.

θ=	35	
b=	3	
p=	0	Sums
$\cos^2(\theta/b^p)$	0.671010072	1.000000000
$\sin^2(\theta/b^p)$	0.328989928	
$\cos^2(\theta/b^{p+1})$	0.959108053	1.000000000
$\sin^2(\theta/b^{p+1})$	0.040891947	
$2 * \cos^2(\theta/b^{p+1})$	1.918216107	
$1 + \cos(\theta/b^p)$	*1.819152044*	
$2 * \sin^2(\theta/b^{p+1})$	0.081783893	
$1 - \cos(\theta/b^p)$	*0.180847956*	

2.10 Wavelets And Multiple-Wavelet Transforms:

The graph in green represents the trigonometric partition graph of the radius as a function of angular velocity. The graph in red represents the X component graph of the trigonometric partitions of the radius as a function of angular velocity. The graph in blue represents the Y component graph of the trigonometric partitions of the radius as a function of angular velocity. In a video on social media, I demonstrate how the theory of chord trigonometric partitions can generate a wide variety of wavelets, both single wavelets and multiple wavelets. This is the easiest tool to generate wavelets and analyze each of the variables and their effects on the graph. The same can be done with other equations of trigonometric partitions of the circular segment, the arrow, the apothem, and the radius growth..., thus causing the family of wavelets to expand. One of the characteristics of this new methodology is that we can easily generate polar functions and analyze these mathematical equations of trigonometric partitions of the chord in the third dimension. This methodology offers us the ease of analyzing each of the nine variables in this equation, where:

k: Torsions or twists.

n: The number of waves, This is one since I have several equations written in Geogebra, which is why it doesn't animate automatically, as it does not cause any effect because this number of waves is a function of the wavelength of the wavelet transform. $n = 2 * \pi/\lambda$.

x_0: Is the position of the wavelet on the X-axis.

ω: Angular velocity, This is one since I have several equations written in Geogebra, which is why it doesn't animate automatically, as it does not cause any effect because these trigonometric partition equations are of the radius as a function of angular velocity $r(\omega)$.

t:Time.

p:Trigonometric partitions of the chord.

r_0: Is the amplitude or initial radius. This can also be applied as the final radius in decrease or compression.

g: Is the position of the multiple or transformed wavelets on the Y-axis.

λ: Wavelength of the wavelet transform.

Another advantage of this system is that with this equation of trigonometric partitions, we can use all the equivalent angle equations $\theta = \omega * t = 2 * \pi * f * t = n * x - \omega * t...$, and we will always be able to generate wavelets. We can even place the equations of the radius partitions as a function of most of the variables of the formula to generate their wavelets. With this trigonometric partition methodology, we will be able to generate the uncertainty principle graph, double slit experiment graph, applying this theory to classical and quantum physics by analyzing all those mathematical equations that have radii in their equations. This new methodology does not require complex numbers since the equation of the radius of the trigonometric partition of the Y component comes to replace complex numbers. Similarly, the Fourier series, SINC trigonometric function or Laplace and Fourier its transforms are not required since these equations were developed as concentric circles like the waves generated by a stone falling into the water, carrying the physical effect of cuts in every sense of a component of a circle as a function of the radii of the trigonometric partitions. In this particular case, if we derive the equation of the trigonometric partitions of the chord of the radius as a function of the angle, we will obtain the chord as a function of the trigonometric partitions. This methodology helps us compare all the equations of the radius of the trigonometric partitions based on each of the variables that can be formed, for example, the radius as a function of velocity $r(\omega)$, the radius as a function of time r(t)... and generate a wide variety of wavelets with their characteristics. The same happens with the equation of the trigonometric partitions of the chord as a function of frequency and time; r(f), r(t). In these equations of this new methodology, it is not required that the functions be periodic as in the case of Fourier series. The functions can be analyzed periodically and non-periodically, without limits on the domain of their values for each variable used in these equations of trigonometric partitions of the chord. The components of the radius as a function of the X coordinate and the Y coordinate are perpendicular, therefore they are orthogonal, and they allow arithmetic properties of addition, subtraction, multiplication, and division for the equations of the X and Y components of the radius of the trigonometric partitions, fulfilling all the conditions of a wavelet. Graphs of the equations of the trigonometric partitions of the Chord as a function of frequency and time, all of these graphs are made with pure mathematical equations in GeoGebra:

$r(x_0) = r_0 * e^{[-2^{p+2} * \sin^2((2*\pi*x_0/\lambda - \omega*t)/2^{p+1})]} + g,$

$r_x(x_0) = r_0 * \cos[k * (2 * \pi * x_0/\lambda - \omega * t)] * e^{[-2^{p+2} * \sin^2((2*\pi*x_0/\lambda - \omega*t)/2^{p+1})]} + g,$

$r_y(x_0) = r_0 * \sin[k * (2 * \pi * x_0/\lambda - \omega * t)] * e^{[-2^{p+2} * \sin^2((2*\pi*x_0/\lambda - \omega*t)/2^{p+1})]} + g,$

Parametric curve:

$a = Curve(r_x(x_0), r_y(x_0), x_0, -2 * \pi, 2 * \pi),$

$r(t) = r_0 * e^{[-2^{p+2} * \sin^2((2*\pi*x_0/\lambda - \omega*t)/2^{p+1})]} + g,$

$r_x(t) = r_0 * \cos[k * (2 * \pi * x_0/\lambda - \omega * t)] * e^{[-2^{p+2} * \sin^2((2*\pi*x_0/\lambda - \omega*t)/2^{p+1})]} + g,$

$r_y(t) = r_0 * \sin[k * (2 * \pi * x_0/\lambda - \omega * t)] * e^{[-2^{p+2} * \sin^2((2*\pi*x_0/\lambda - \omega*t)/2^{p+1})]} + g,$

Parametric curve:

$a = Curve(r_x(t), r_y(t), t, -2 * \pi, 2 * \pi),$

$r(\omega) = r_0 * e^{[-2^{p+2} * \sin^2((2*\pi*x_0/\lambda - \omega*t)/2^{p+1})]} + g,$

$r_x(\omega) = r_0 * \cos[k * (2 * \pi * x_0/\lambda - \omega * t)] * e^{[-2^{p+2} * \sin^2((2*\pi*x_0/\lambda - \omega*t)/2^{p+1})]} + g,$

$r_y(\omega) = r_0 * \sin[k * (2 * \pi * x_0/\lambda - \omega * t)] * e^{[-2^{p+2} * \sin^2((2*\pi*x_0/\lambda - \omega*t)/2^{p+1})]} + g,$

Parametric curve:

$a = Curve(r_x(\omega), r_y(\omega), \omega, -2 * \pi, 2 * \pi).$

Figure 2.33: Transforms Of The Trigonometric Partitions Of The Chord.

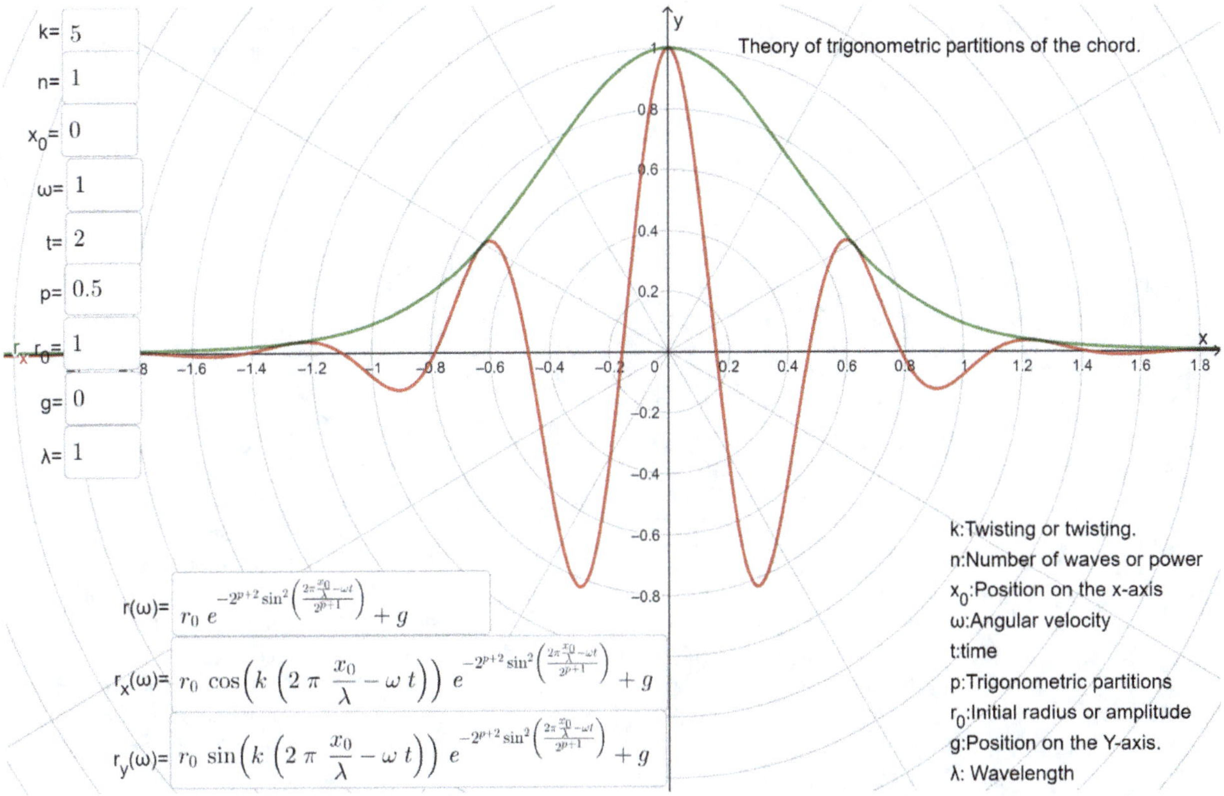

Note:Wavelet Charts with Trigonometric Partitions, x component of the chord.

Figure 2.34: Transforms Of The Trigonometric Partitions Of The Chord.

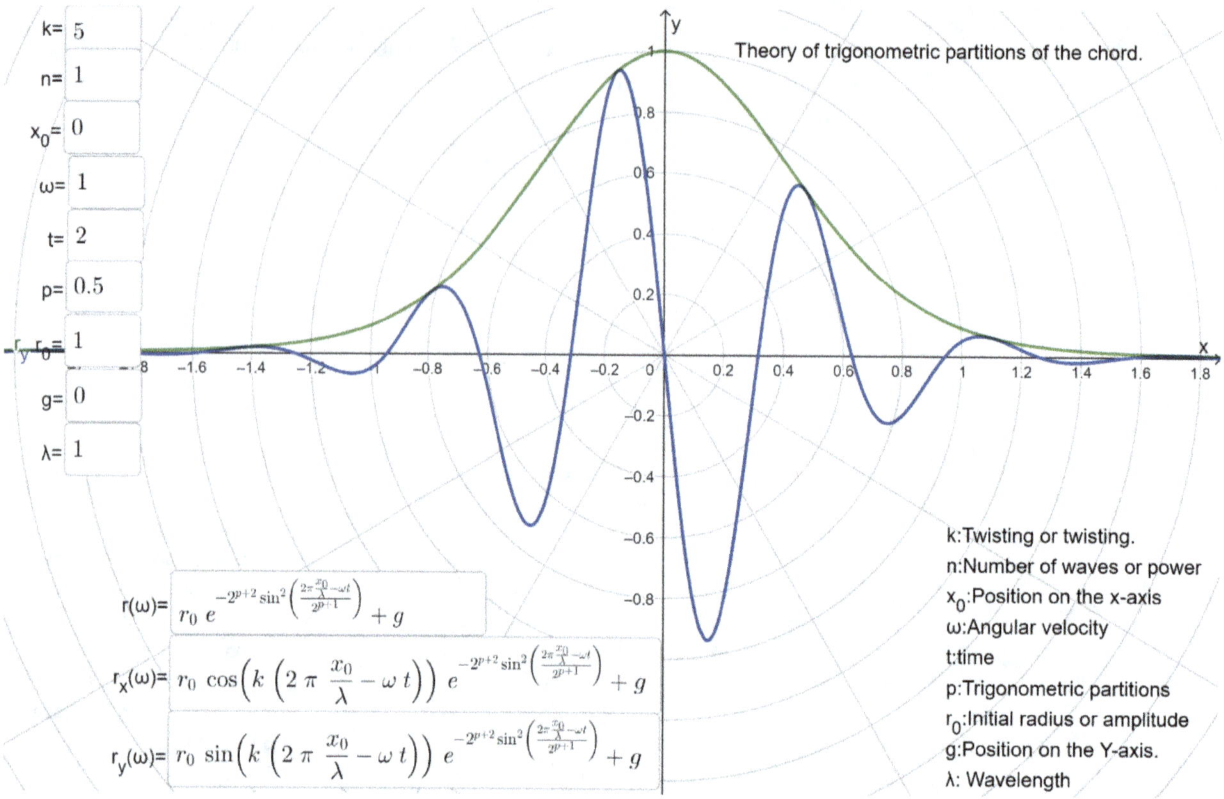

Note:Wavelet Charts with Trigonometric Partitions, y component of the chord.

Figure 2.35: Transforms Of The Trigonometric Partitions Of The Chord.

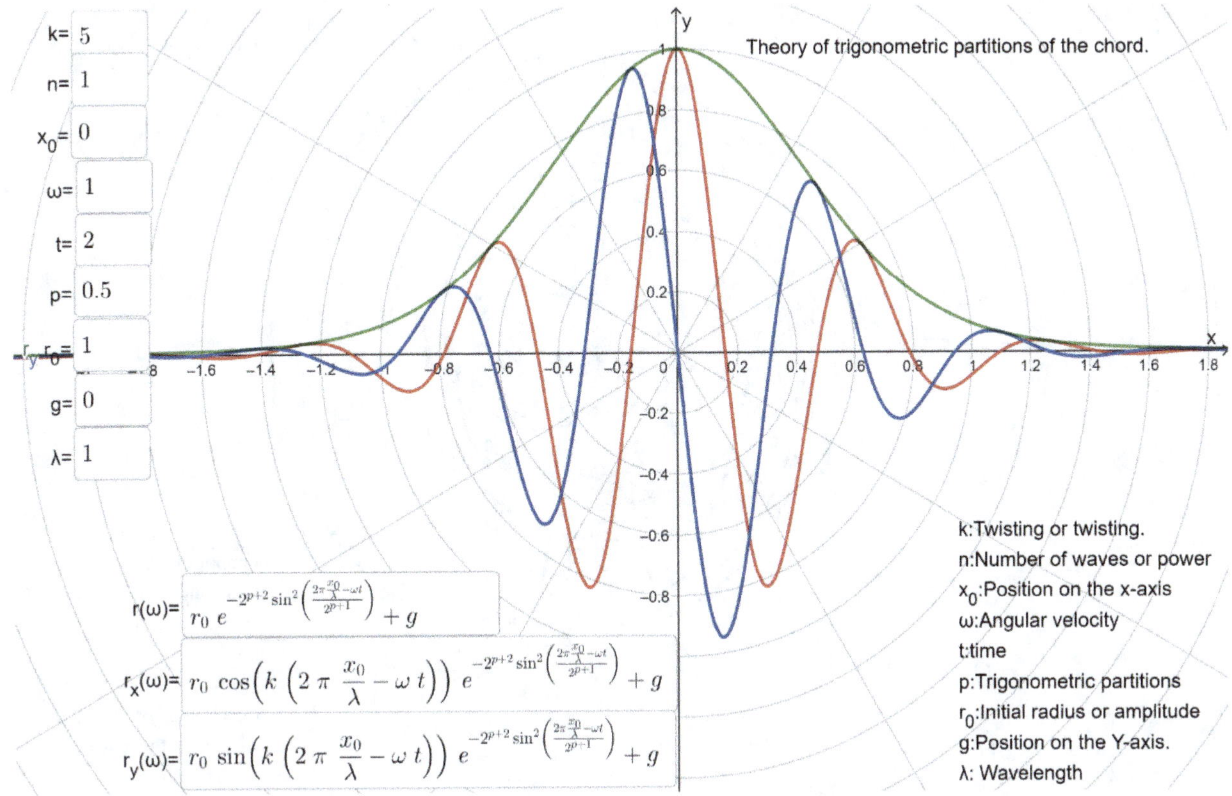

Note:Wavelet Charts with Trigonometric Partitions, x and y components of the chord.

Figure 2.36: Transforms Of The Trigonometric Partitions Of The Chord.

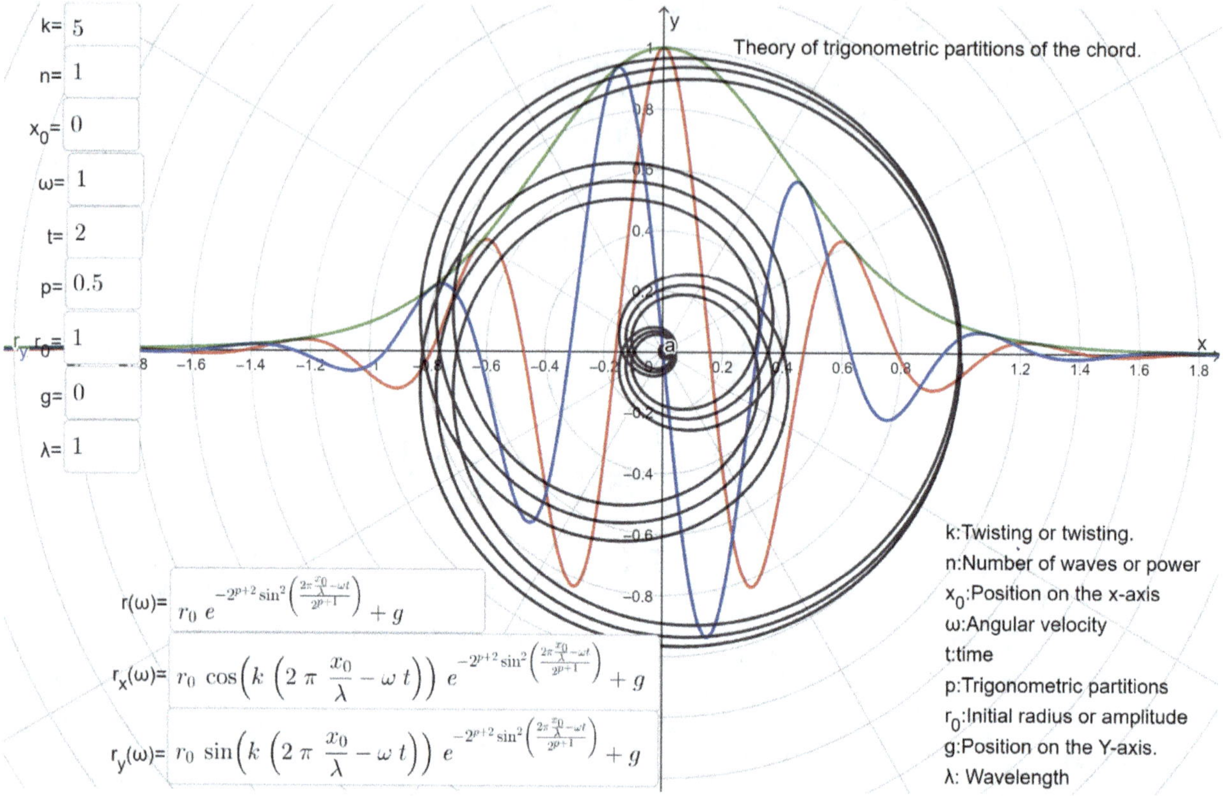

Note:Wavelet Charts with Trigonometric Partitions, x,y and parametric curve components of the chord.

Figure 2.37: Transforms Of The Trigonometric Partitions Of The Chord.

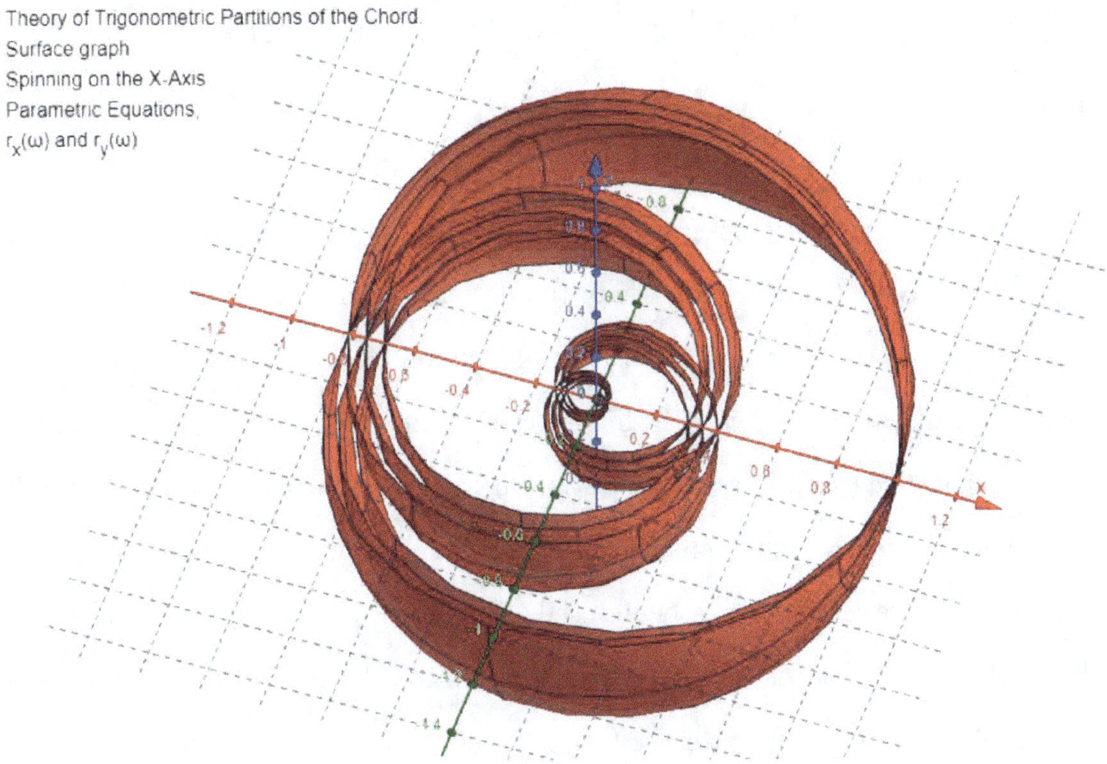

Note:Wavelet Charts with Trigonometric Partitions of the chord, Surface graph, Spinning on the X-Axis.

2.11 Multiwavelets Transforms As A Function Of Frequency, Angular Velocity, And Time.

If we ask ourselves: How are wavelets generated? These equations of trigonometric chord partitions come from concentric circular analysis, similar to the waves generated in water when a stone falls. With them, we can study the decrease of the final radius, or rearrange the original formula of the chord partition to study the initial radius and analyze its birth or the growth of the initial radius, which is why the exponent 2 is negative. With the Theory of Trigonometric Partitions, we can easily generate and analyze the wavelets and their transforms.There are several equations of trigonometric partitions; in this particular case, the Trigonometric Chord Partition is being used. With this new concept, we can perform 2D analysis with its parametric curves of one equation for the **X** coordinate and another for the **Y** coordinate, also in 3D with the technique of solid revolution which is equivalent to performing an integral. In these equations of the radius of trigonometric partitions, we will use the equivalent equations of the angle as a function of frequency and time to generate multiple wavelets, likewise with the equivalent equation of the angle as a function of the constant **k**; which in this case is equal to the rotations, twist, and wave numbers, the position **x**, of the Angular Velocity ω, and time **t**; various wavelets can be generated. By studying each of the characteristics, for example, if we increase the frequency, we observe how the wavelet compresses, observing the effects that each variable causes in the mathematical equations. In these equations of trigonometric partitions of the wavelets, we can write the radius as a function of frequency, and the radius as a function of time, and we will always generate wavelets with their characteristics. Therefore, if we write these equations of the radius as a function of trigonometric partitions **p**, and the constant **k**, wavelet transforms are not generated. Similarly, in the equation as a function of position and time, we can use all the variables to write the radius as a function of each of them, and we will always generate wavelets. With the first derivative of the trigonometric partitions of the chord as a function of the trigonometric partitions **p**, we can generate scalograms. This same new methodology can be used in the equations of trigonometric partitions of the circular sector, apothem, arrow, and radius growth, creating a wide variety of wavelets and their transforms. This is a new technique that does not require software for its programming, complex numbers, or analysis with the Fourier series or its transforms, the tool used is GeoGebra, initially, I started making them in Excel spreadsheet. With these same equations, we can create the graph of the uncertainty principle; in the graphs generated by this theory, you will observe the distortion caused by the harmonics in the sine wave, and the DC offset of the equation of the **Y** coordinate. This knowledge can be applied to vibration analysis... etc. If you want to learn more about the Theory of Trigonometric Partitions, you should read the book I have written on this and other topics of its application. Wavelets as a function of frequency, angular velocity, and time. With these equations of trigonometric chord partitions of the radius as a function of time, it can be compared with the Fast Fourier Transform FFT. The equation of trigonometric chord partitions of the radius as a function of angular velocity can be compared with the Discrete Fourier Transform DWT. With these new equations of trigonometric partitions, it will be possible to study quantum entanglement, quantum tunneling. With this new technique, we can use integers for the variables of the equation and create wavelets. Even if we increase the power of this equation of trigonometric partitions, wavelets and the graph of the uncertainty principle can continue to be generated. This is pure mathematics, and with these equations of trigonometric partitions, we can apply arithmetic properties and create wavelets. The GeoGebra sliders are the domain of each variable, and in these equations, they have no limits. Any value from negative infinity to positive infinity can be used. With this methodology, wavelets generated at low and high frequencies can be analyzed. The Theory of Trigonometric Partitions is a super-symmetric equation; it can behave like a dot, a straight line, or a curve, and works in the real plane as well as in the complex number plane. Graphs of the equations of the trigonometric partitions of the Chord as a function of frequency and time, all of these graphs are made with pure mathematical equations in GeoGebra:

$R_2(t) = r_0^n * e^{[-n*2^{p+2}*\sin^2(\pi*f*t/2^p)]},$

$R_{2x}(t) = r_0^n * e^{[-n*2^{p+2}*\sin^2(\pi*f*t/2^p)]} * [cos(2*\pi*k*f*t)]^n,$

$R_{2y}(t) = r_0^n * e^{[-n*2^{p+2}*\sin^2(\pi*f*t/2^p)]} * [sin(2*\pi*k*f*t)]^n,$

Parametric curve:

$e = Curve(R_{2x}(t), R_{2y}(t), t, -2*\pi, 2*\pi),$

$R_2(f) = r_0^n * e^{[-n*2^{p+2}*\sin^2(\pi*f*t/2^p)]},$

$R_{2x}(f) = r_0^n * e^{[-n*2^{p+2}*\sin^2(\pi*f*t/2^p)]} * [cos(2*\pi*k*f*t)]^n,$

$R_{2y}(f) = r_0^n * e^{[-n*2^{p+2}*\sin^2(\pi*f*t/2^p)]} * [sin(2*\pi*k*f*t)]^n,$

Parametric curve:

$e = Curve(R_{2x}(f), R_{2y}(f), f, -2*\pi, 2*\pi).$

Figure 2.38: Transforms Of The Trigonometric Partitions Of The Chord.

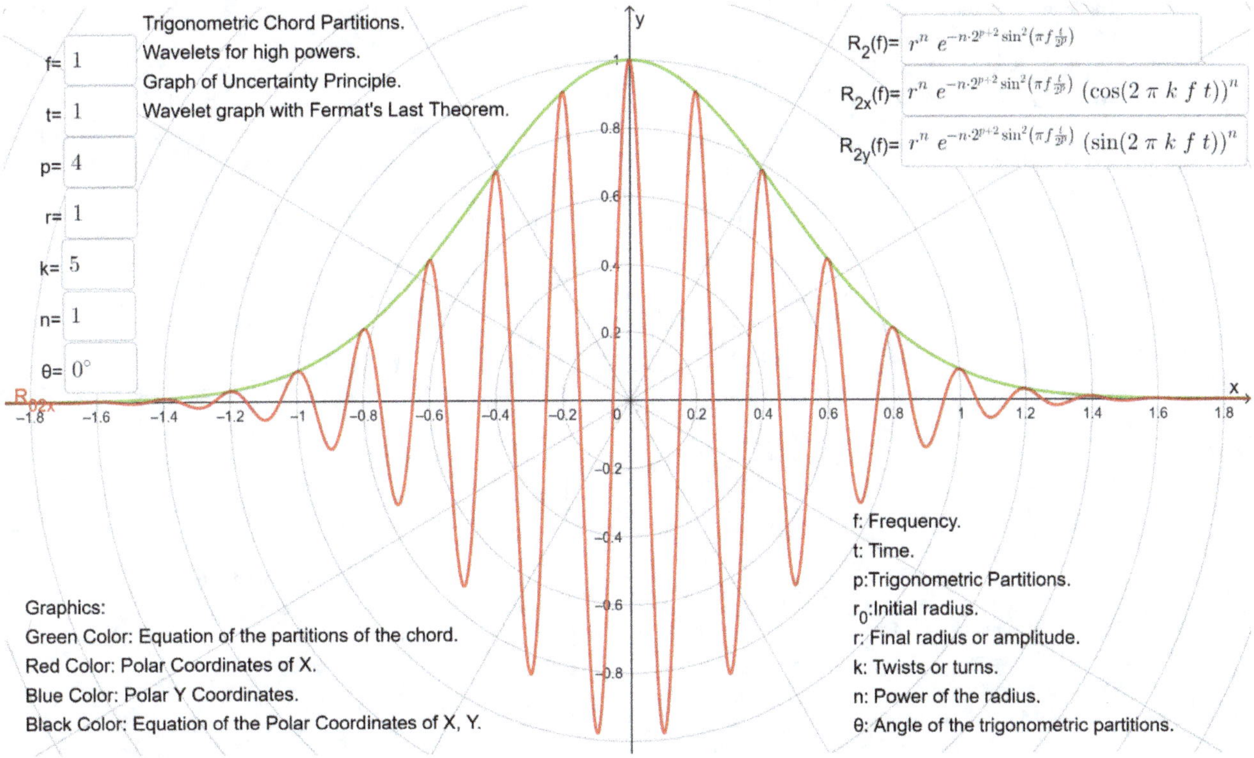

Note:Wavelet Charts with Trigonometric Partitions, x component of the chord.

Figure 2.39: Transforms Of The Trigonometric Partitions Of The Chord.

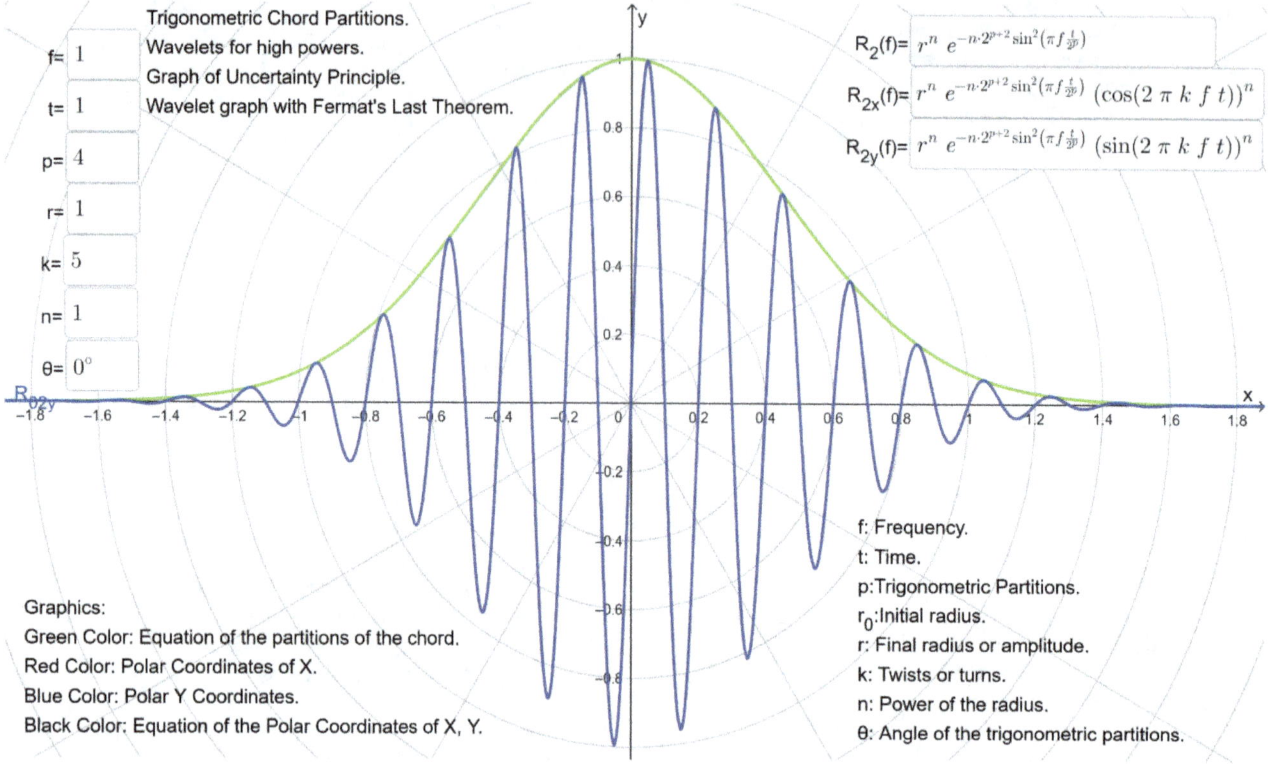

Note:Wavelet Charts with Trigonometric Partitions, y component of the chord.

Figure 2.40: Transforms Of The Trigonometric Partitions Of The Chord.

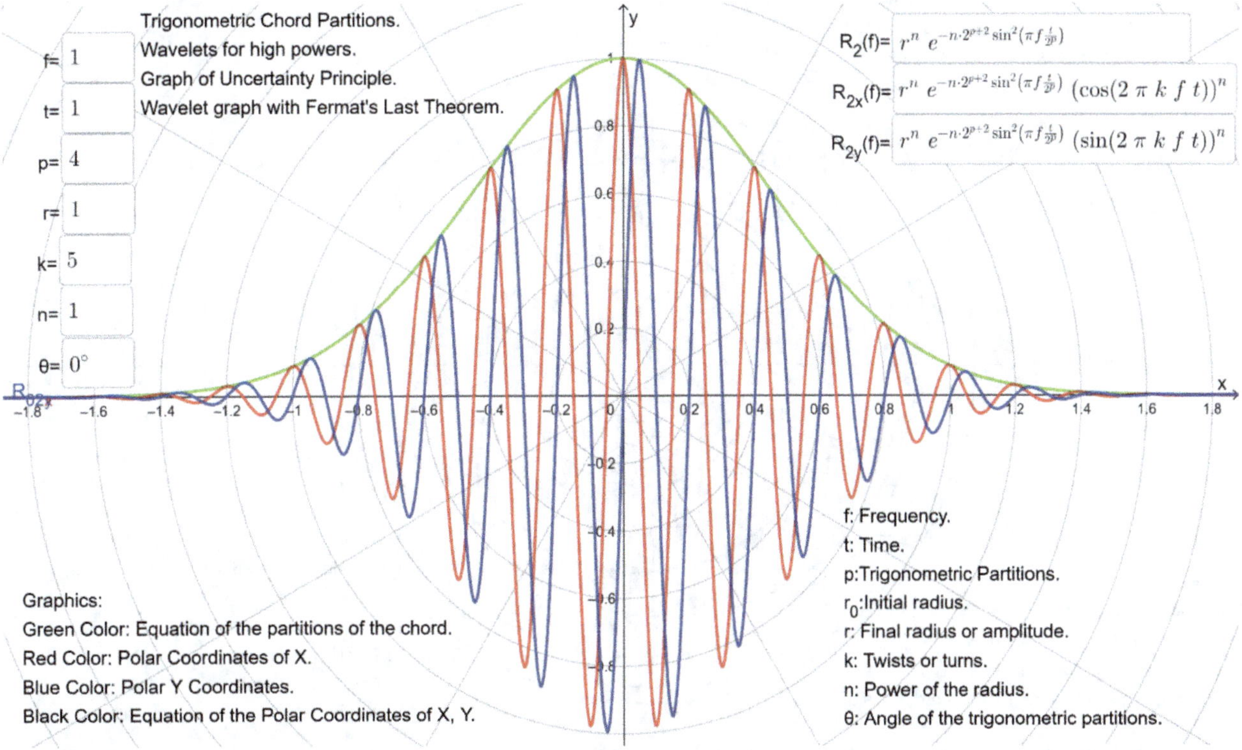

Note:Wavelet Charts with Trigonometric Partitions, x and y components of the chord.

Figure 2.41: Transforms Of The Trigonometric Partitions Of The Chord.

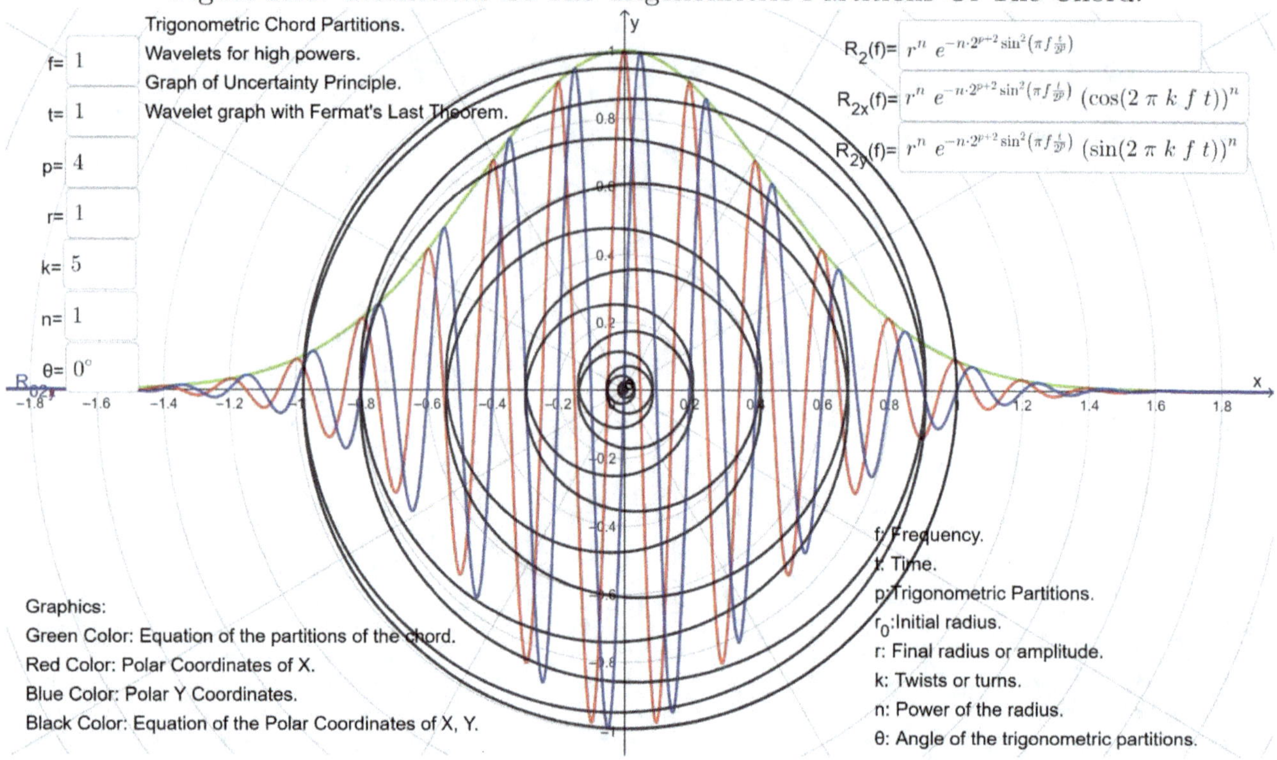

Note: Wavelet Charts with Trigonometric Partitions, x,y and parametric curve components of the chord.

Figure 2.42: Transforms Of The Trigonometric Partitions Of The Chord.

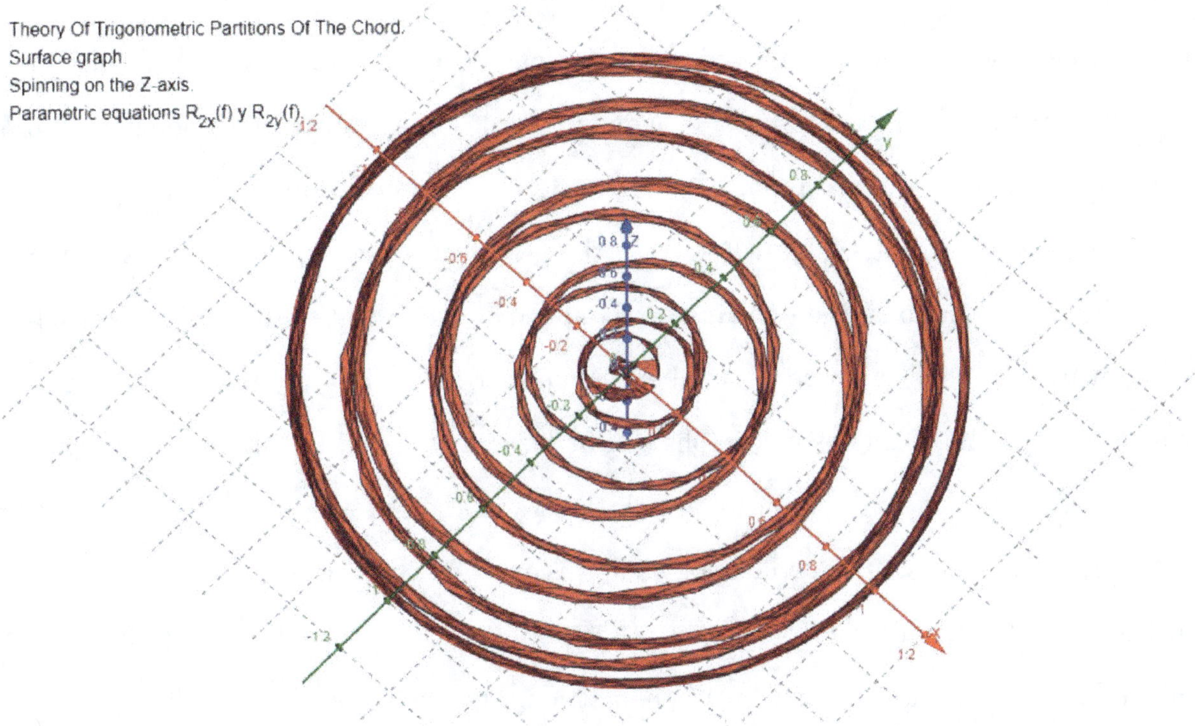

Note:Wavelet Charts with Trigonometric Partitions of the chord, Surface graph, Spinning on the Z-Axis.

2.12 Wavelets Of nth Powers:

These wavelets of nth powers are based on the Pythagorean equation $r^2 = x^2 + y^2$, although I don't sum the components x^2 and y^2. Therefore, each equation of the trigonometric chord partitions is raised to the nth power, similar to Fermat's Last Theorem. Similarly, a wide variety of wavelets can be generated, such as the graph generated in the double-slit experiment. In these equations, the domains are bounded to small values to prevent GeoGebra from slowing down due to the large number of calculations used by the program. When increased to large values, interesting graphs can be observed. When we change each variable, we can analyze the effects that each variable causes in the equations of the trigonometric chord partitions. This simple methodology helps us analyze physical cases like the double-slit experiment. Additionally, multiple graphs of the double-slit experiment and wavelets can be generated. This new methodology of trigonometric partitions serves us to study wavelets at both low and high frequencies. In the book I have written, you can learn about the various applications where the theory of trigonometric partitions can be used. This tool is similar to the Fourier series theory, but it is not the Fourier series. With this new partition tool, the number of known wavelets has increased. Initially, when I started developing this theory, I applied the concepts of spiral angles to the double-slit experiment, where I had successful results. Later, I realized that I had radii and could also apply trigonometric partitions. In a social media video, I recorded, observe what happens when prime and odd numbers are used and how the wavelet shape changes. Additionally, with the Theory of Trigonometric Partitions, we can study the harmonics that deform the sine and cosine waves and generate their graphs. With the first and second derivatives of the trigonometric partitions of the chord can also generate duplex or double wavelets. Graphs of the equations of the trigonometric partitions of the chord as a function of frequency and time:

The graph generated by the double-slit experiment is a wavelet of power two.

$R_2(t) = r_0^n * e^{[-n*2^{p+2}*\sin^2(\pi*f*t/2^p)]}$,

$R_{2x}(t) = r_0^n * e^{[-n*2^{p+2}*\sin^2(\pi*f*t/2^p)]} * [cos(2*\pi*k*f*t)]^n$,

$R_{2y}(t) = r_0^n * e^{[-n*2^{p+2}*\sin^2(\pi*f*t/2^p)]} * [sin(2*\pi*k*f*t)]^n$,

Parametric curve:

$e = Curve(R_{2x}(t), R_{2y}(t), t, -2*\pi, 2*\pi)$,

$R_2(f) = r_0^n * e^{[-n*2^{p+2}*\sin^2(\pi*f*t/2^p)]}$,

$R_{2x}(f) = r_0^n * e^{[-n*2^{p+2}*\sin^2(\pi*f*t/2^p)]} * [cos(2*\pi*k*f*t)]^n$,

$R_{2y}(f) = r_0^n * e^{[-n*2^{p+2}*\sin^2(\pi*f*t/2^p)]} * [sin(2*\pi*k*f*t)]^n$,

Parametric curve:

$e = Curve(R_{2x}(f), R_{2y}(f), f, -2*\pi, 2*\pi)$,

A:This graph is for the light:X Coordinate.

B:This graph is for the darkness:Y Coordinate.

C and D:Charts combination: X and Y Coordinate.

E:Surface graph: The red line is the parametric curve.

Figure 2.43: Transforms Of The Trigonometric Partitions Of The Chord.

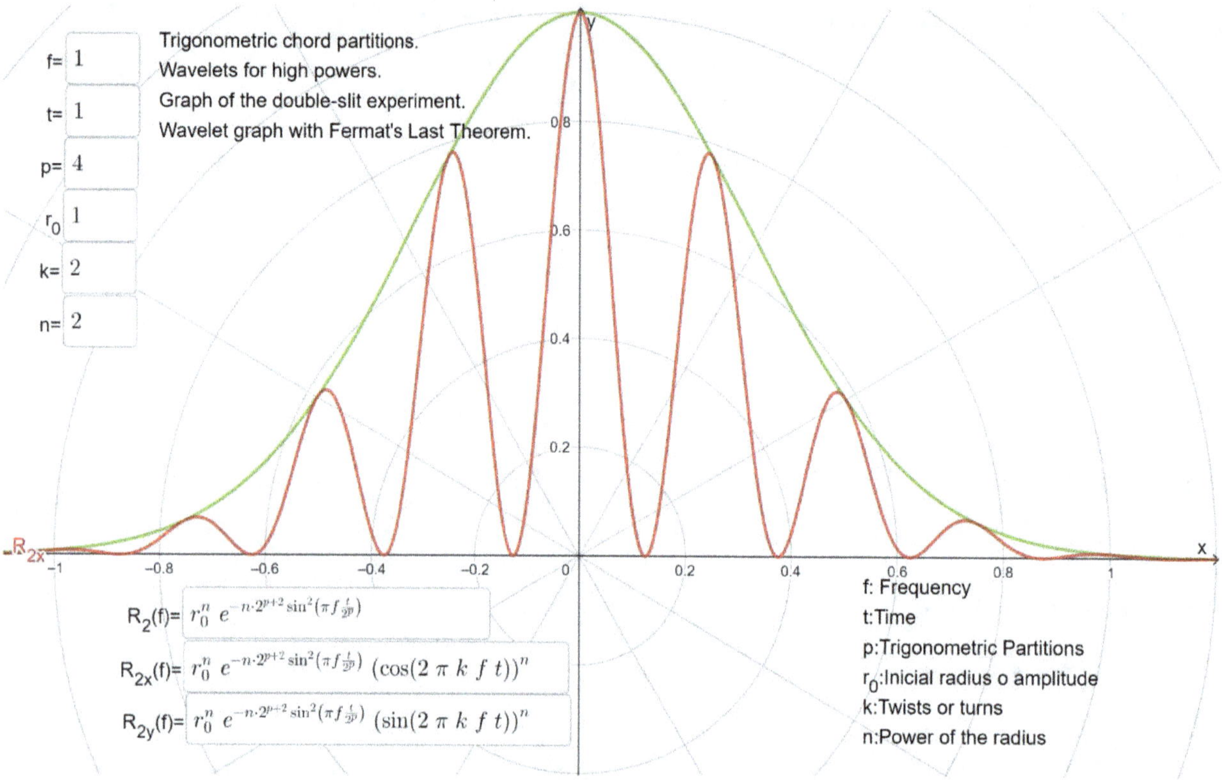

Note:Wavelet Charts with Trigonometric Partitions, x component of the chord.

Figure 2.44: Transforms Of The Trigonometric Partitions Of The Chord.

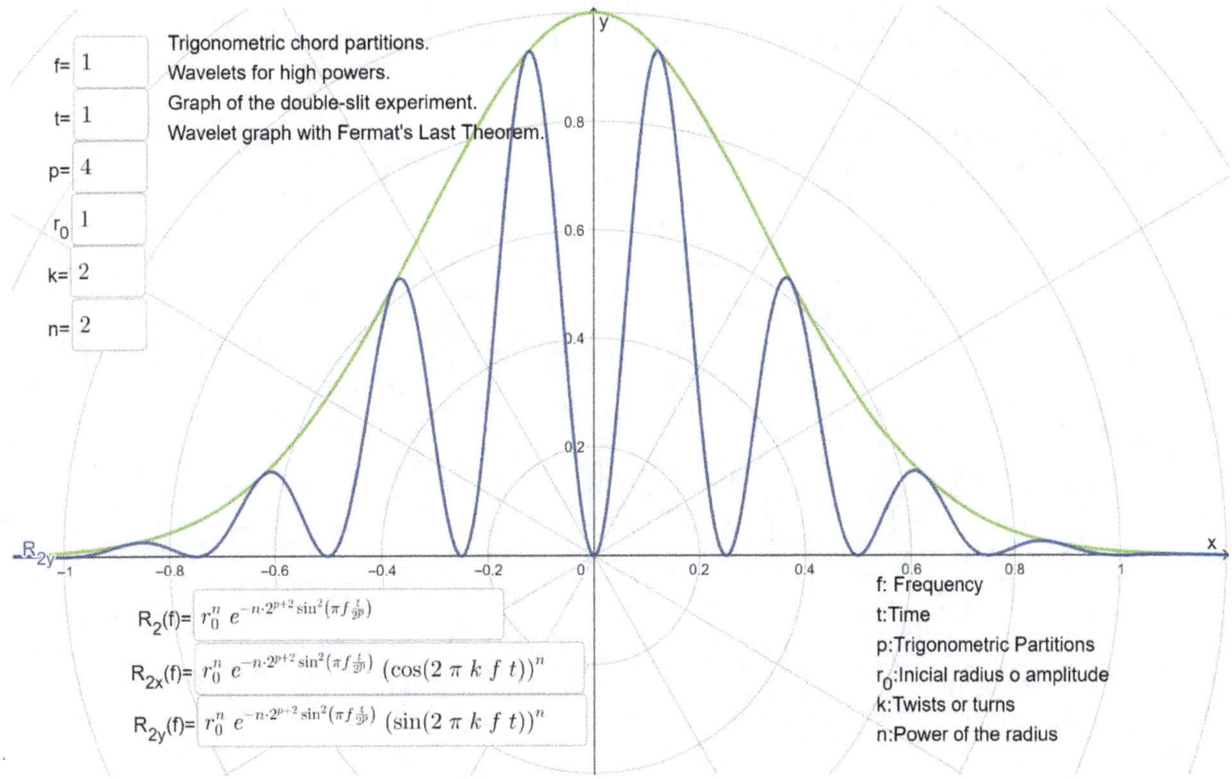

Note:Wavelet Charts with Trigonometric Partitions, y component of the chord.

Figure 2.45: Transforms Of The Trigonometric Partitions Of The Chord.

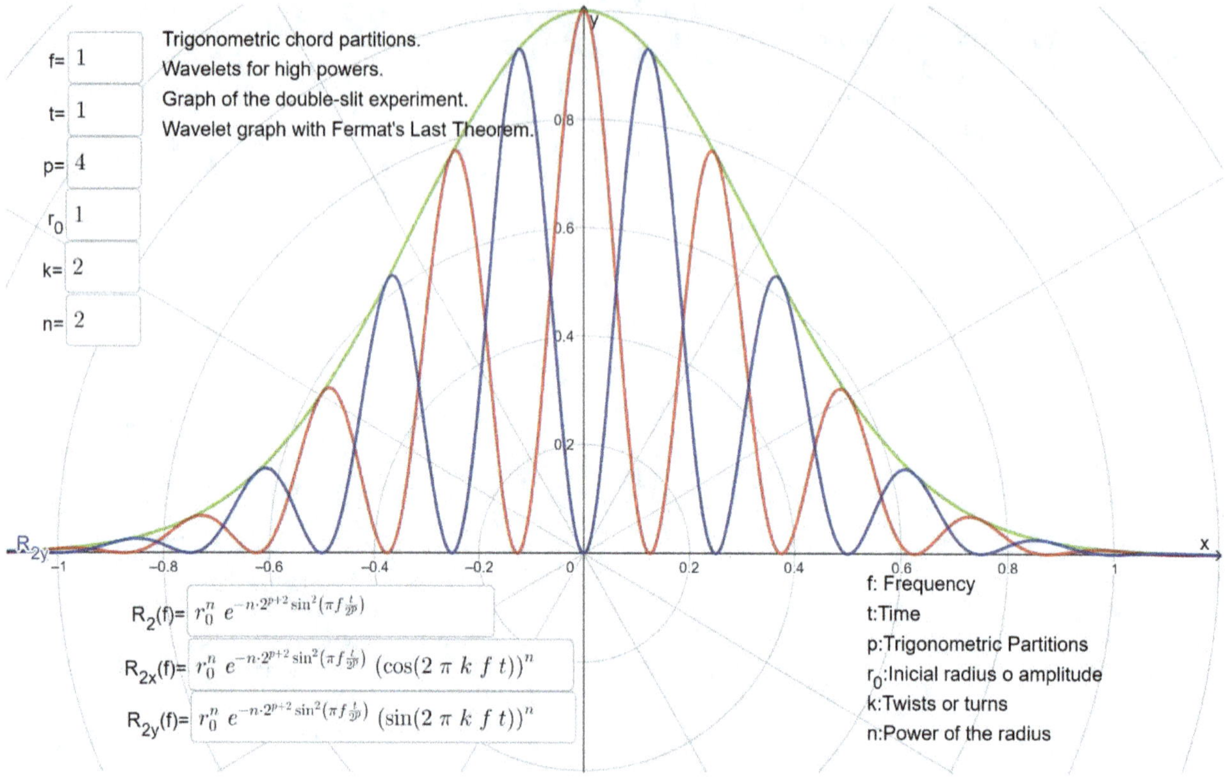

Note: Wavelet Charts with Trigonometric Partitions, x and y components of the chord.

Figure 2.46: Transforms Of The Trigonometric Partitions Of The Chord.

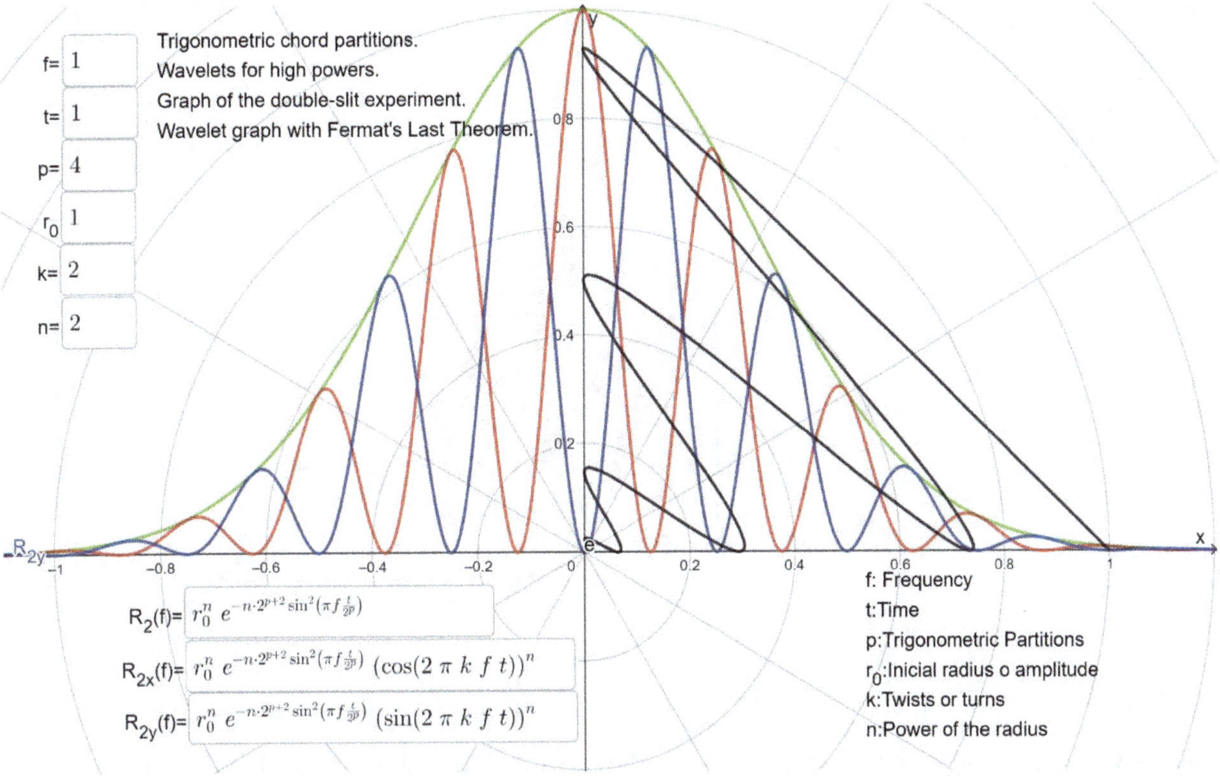

Note:Wavelet Charts with Trigonometric Partitions, x,y and parametric curve components of the chord.

Figure 2.47: Transforms Of The Trigonometric Partitions Of The Chord.

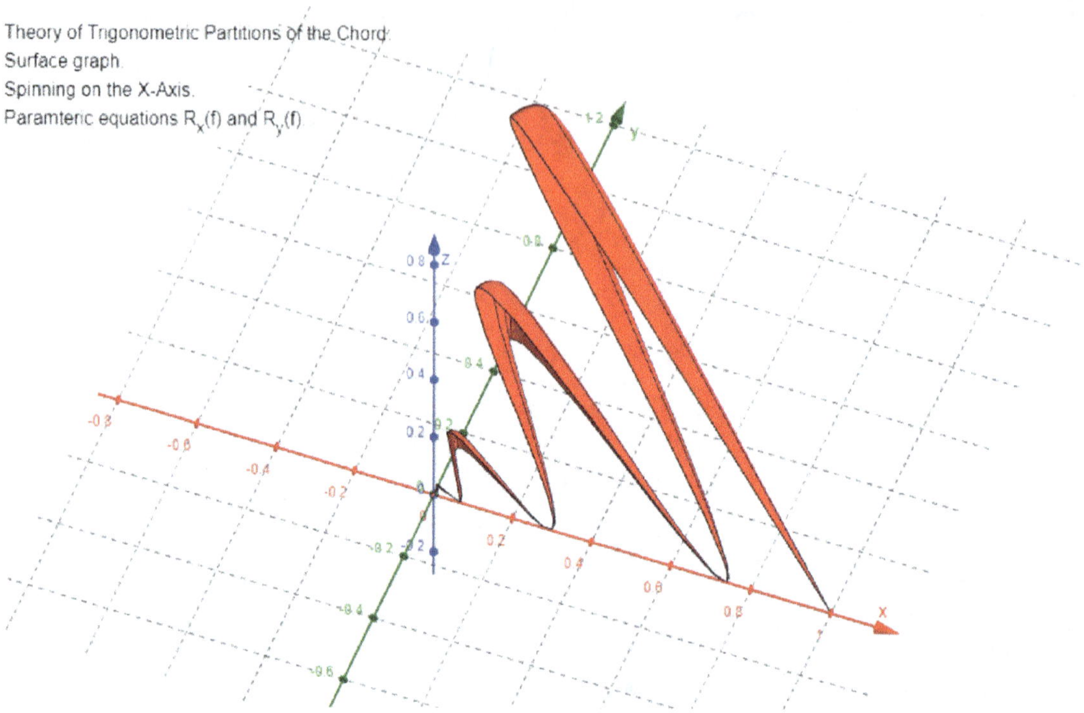

Note:Wavelet Charts with Trigonometric Partitions of the chord, Surface graph, Spinning on the X-Axis.

I have written a book about the Spiral Angles, Trigonometric Spirals and Trigonometric Partitions. You can buy and learn all about the application of this new method. Visit my social networks where I talk more about this topic of the Theory of Spiral Angles, Spirals, and Trigonometric Partitions. In these teachings where I talk about some topics that are not written and explained in the book.

2.13 Wavelets With Damped Simple Harmonic Motion

The force exerted by a spring, commonly known as Hooke's law, is equal to the spring constant multiplied by its displacement along the **x**–axis or **y**–axis. Its mathematical expression in the **x**–direction is $F = -cx$, and in the **y**–direction is $F = -cy$, where $\omega = \sqrt{c/m}$. This can also be expressed as a second order differential equation using Newton's second law. By drawing an analogy with the mathematical equations of the trigonometric partitions, in order to continue generating wavelets, two types of comparative analyses can be carried out with all the equations of the trigonometric partitions.

First: if we take the trigonometric partition equation of the chord:

$$r_0(t) = r\, e^{-2^{p+2}\sin^2(\omega t/2^{p+1})},$$

considering the amplitude variable **r** as a constant, and compute the derivative of the coordinate r_0 with respect to time, $dr_0/dt = r_0' = V_{0r}$, and subsequently obtain its **x** and **y** components by multiplying by the cosine and sine as functions of the angular velocity and time, scaled by the constant **k** which produces the wavelets, we obtain $dr_{0x}/dt = r_{0x}' = V_{r0x}$ and $dr_{0y}/dt = r_{0y}' = V_{r0y}$.

Second: if we take the trigonometric partition equation of the chord:

$$r_0(t) = r\, e^{-2^{p+2}\sin^2(\omega t/2^{p+1})},$$

again considering the amplitude variable **r** as a constant, with the components

$$r_{0x}(t) = r\,\cos(k\omega t)\, e^{-2^{p+2}\sin^2(\omega t/2^{p+1})} \quad \text{and} \quad r_{0y}(t) = r\,\sin(k\omega t)\, e^{-2^{p+2}\sin^2(\omega t/2^{p+1})},$$

then the derivatives of the initial radius coordinates with respect to time that is, the radial velocities of the **x** and **y** coordinates of the initial radius are given by:

$dr_{0x}/dt = r'_{0x} = V_{r0x}$ and $dr_{0y}/dt = r'_{0y} = V_{r0y}$.

The first derivatives of the velocity components $\mathbf{x}$ and $\mathbf{y}$ are:

$$V_{r0x} = -r * \omega\,[\,k \sin(k\omega t) + 2\cos(k\omega t)\,\sin(\omega t/2^p)\,]\,e^{-2^{p+2}\sin^2(\omega t/2^{p+1})}$$

$$V_{r0y} = -r * \omega\,[\,-k \cos(k\omega t) + 2\sin(k\omega t)\,\sin(\omega t/2^p)\,]\,e^{-2^{p+2}\sin^2(\omega t/2^{p+1})}$$

If we proceed with the second derivative a longer and more complex process we obtain the accelerations for both coordinates, which can then be used in the second order differential equation.

This same first and second procedure can be applied to all trigonometric partition equations in order to continue generating wavelets.

Figure 2.48: First derivative of the initial radius with respect to time for the chord.

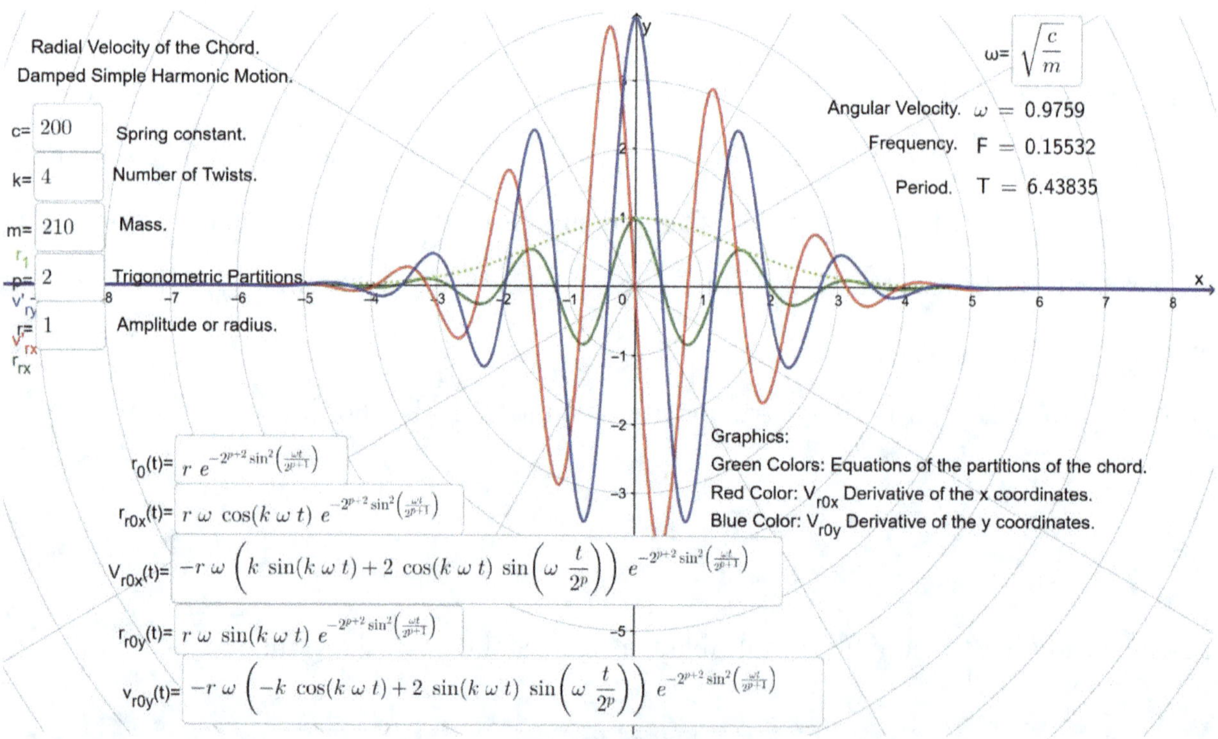

Note: This figure shows how wavelets are generated by the application of trigonometric partitions to damped simple harmonic motion, using the radial velocity of the trigonometric partitions.

With the trigonometric partition equation of the apothem which is a simpler and shorter expression to differentiate in its first and second derivatives for its **x** and **y** components given by

$$r_0(t) = r\, e^{-2^p \sin(\omega t/2^p)},$$

considering the amplitude variable **r** as a constant, its components are:

$$r_{0x}(t) = r\, \cos(k\omega t)\, e^{-2^p \sin(\omega t/2^p)} \quad \text{and} \quad r_{0y}(t) = r\, \sin(k\omega t)\, e^{-2^p \sin(\omega t/2^p)}.$$

To obtain the derivatives of the initial radius coordinates with respect to time which correspond to the radial velocities of the **x** and **y** coordinates of the initial radius we have:

$$\frac{dr_{0x}}{dt} = r'_{0x} = V_{r0x} \quad \text{and} \quad \frac{dr_{0y}}{dt} = r'_{0y} = V_{r0y}.$$

$$\frac{d^2r_{0x}}{dt^2} = r''_{0x} = a_{r0x} \quad \text{and} \quad \frac{d^2r_{0y}}{dt^2} = r''_{0y} = a_{r0y}.$$

The first derivatives for the **x** and **y** velocity components are:

$$V_{r0x} = -r\,\omega \left[k \sin(k\omega t) + \cos(k\omega t)\, \cos(\omega t/2^p)\right] e^{-2^p \sin(\omega t/2^p)}$$

$$V_{r0y} = -r\,\omega \left[-k \cos(k\omega t) + \sin(k\omega t)\, \cos(\omega t/2^p)\right] e^{-2^p \sin(\omega t/2^p)}.$$

Figure 2.49: First derivative of the initial radius with respect to time for the apothem.

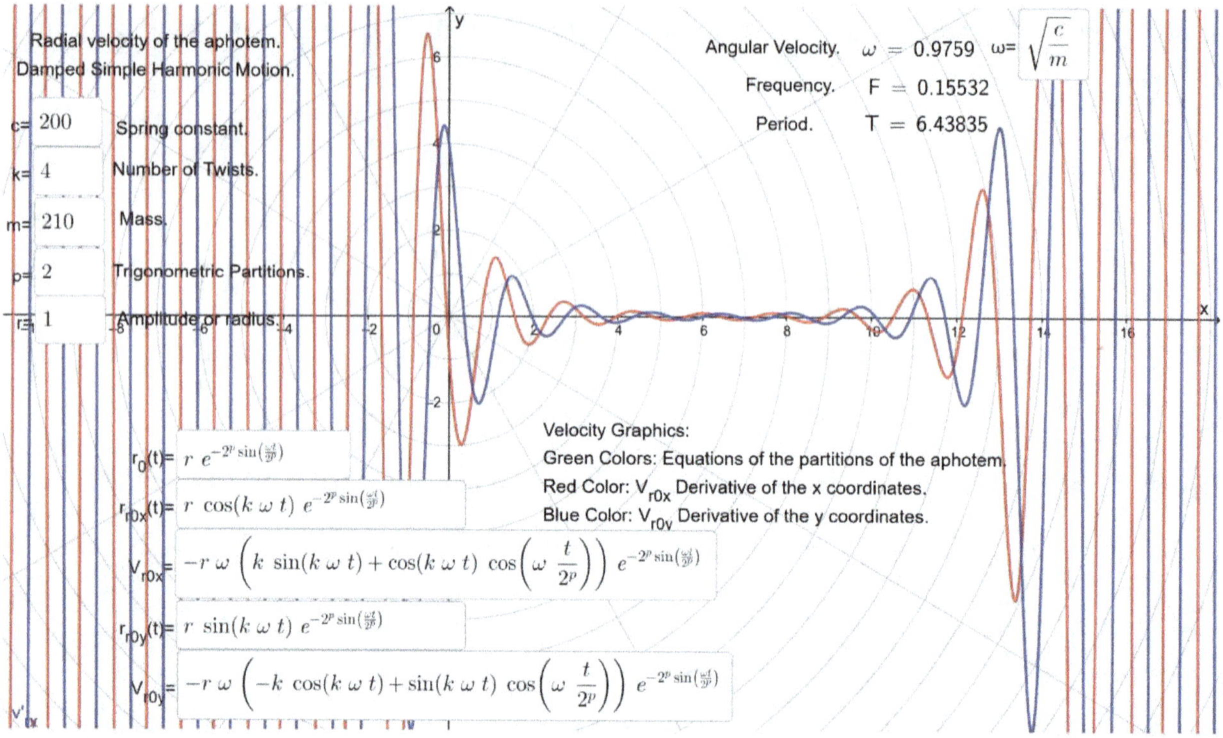

Note: This figure illustrates how large wavelets are generated by the application of trigonometric partitions to damped simple harmonic motion, through the radial velocity of the trigonometric partitions.

Figure 2.50: First derivative of the initial radius with respect to time for the radial growth.

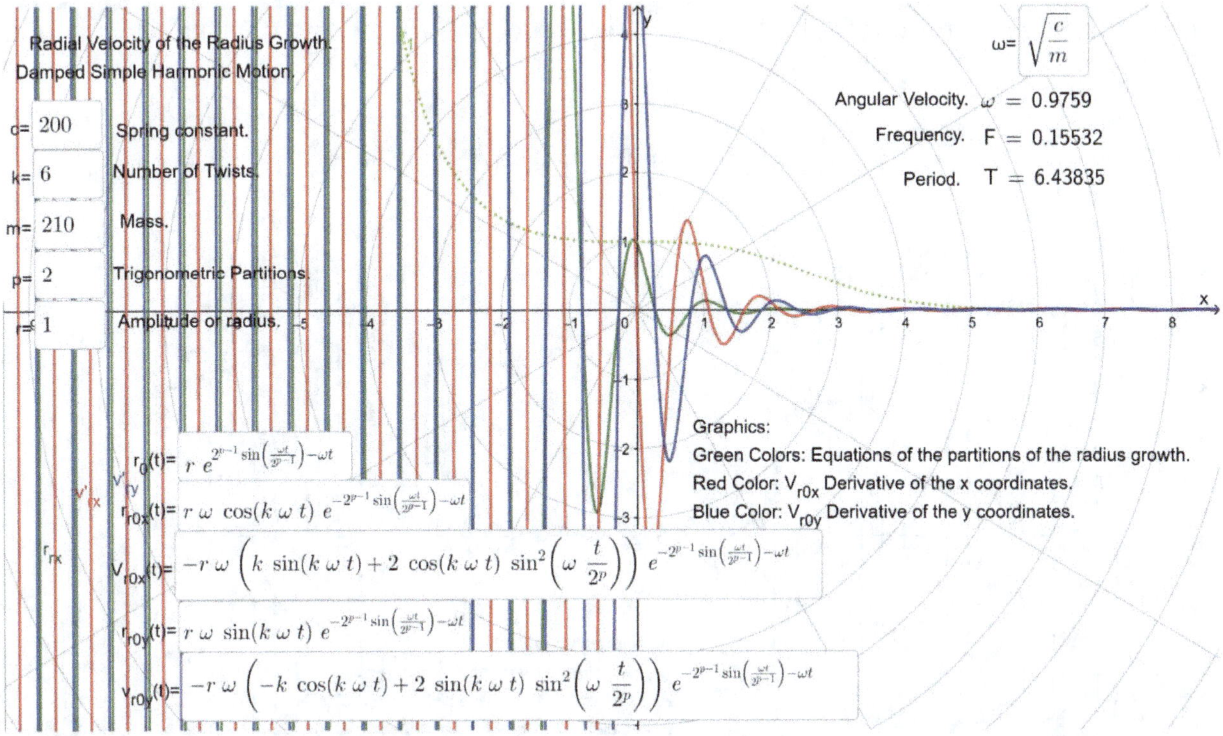

Note: This figure shows how wavelets are generated by the application of trigonometric partitions to damped simple harmonic motion, using the radial velocity of the trigonometric partitions.

If we continue by computing the second derivative a longer and more complex process we obtain the accelerations for both coordinates, which can then be used in the second order differential equation.

Figure 2.51: Second derivative of the initial radius with respect to time for the apothem.

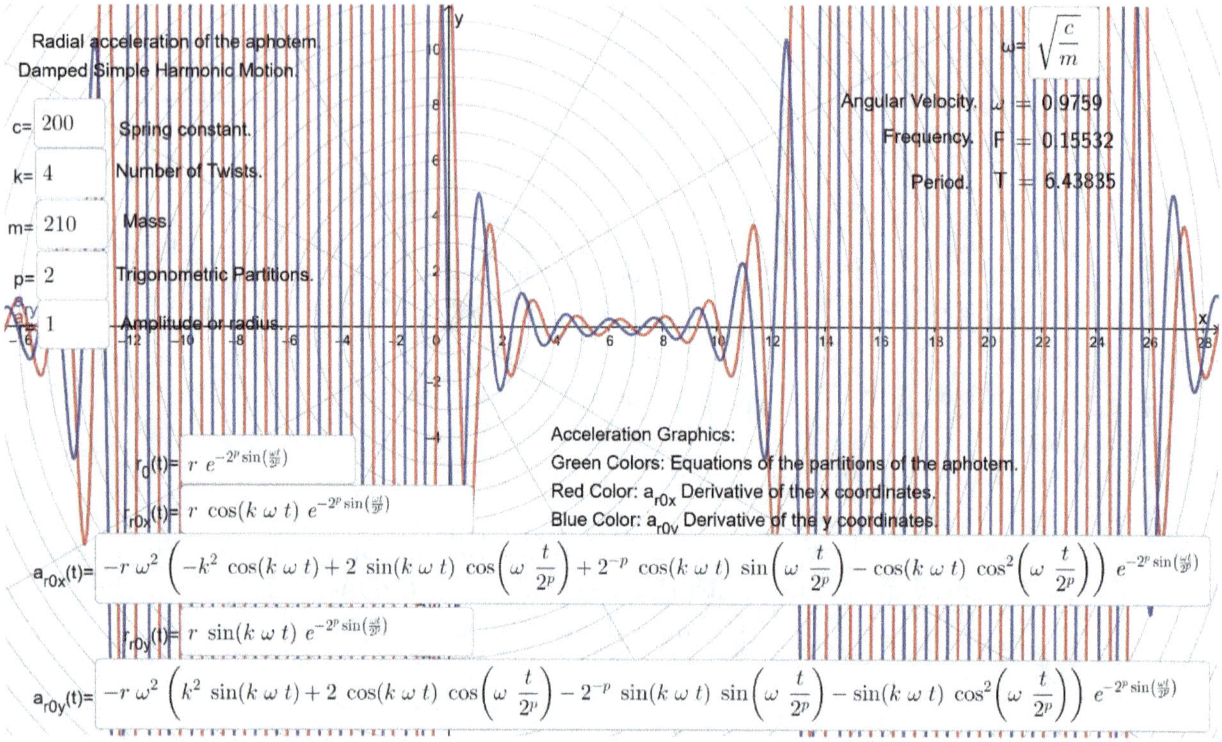

$$\omega = \sqrt{\frac{c}{m}}$$

$$r_d(t) = r\, e^{-2^p \sin\left(\frac{\omega t}{2^p}\right)}$$

$$r_{r0x}(t) = r \cos(k\,\omega\,t)\, e^{-2^p \sin\left(\frac{\omega t}{2^p}\right)}$$

$$a_{r0x}(t) = -r\,\omega^2 \left(-k^2 \cos(k\,\omega\,t) + 2\sin(k\,\omega\,t)\cos\left(\omega\,\frac{t}{2^p}\right) + 2^{-p}\cos(k\,\omega\,t)\sin\left(\omega\,\frac{t}{2^p}\right) - \cos(k\,\omega\,t)\cos^2\left(\omega\,\frac{t}{2^p}\right)\right) e^{-2^p \sin\left(\frac{\omega t}{2^p}\right)}$$

$$r_{r0y}(t) = r \sin(k\,\omega\,t)\, e^{-2^p \sin\left(\frac{\omega t}{2^p}\right)}$$

$$a_{r0y}(t) = -r\,\omega^2 \left(k^2 \sin(k\,\omega\,t) + 2\cos(k\,\omega\,t)\cos\left(\omega\,\frac{t}{2^p}\right) - 2^{-p}\sin(k\,\omega\,t)\sin\left(\omega\,\frac{t}{2^p}\right) - \sin(k\,\omega\,t)\cos^2\left(\omega\,\frac{t}{2^p}\right)\right) e^{-2^p \sin\left(\frac{\omega t}{2^p}\right)}$$

Note: This figure illustrates how large wavelets are generated by the application of trigonometric partitions to damped simple harmonic motion, through the radial velocity of the trigonometric partitions.

All the equations of the Trigonometric Partitions are useful for analyzing the relationship that exists within the numerical system, particularly the relationship between irrational numbers and complex numbers, which are not easily interpreted in either the real or complex plane. The following diagram helps to establish this relationship, since the equations of the trigonometric partitions can operate in any form within a circle, whether in its polar form, and follow equation or number patterns, depending entirely on which equations are used for the angle, the radius, or the trigonometric partitions **p**. For example, if we use the equation of the radius derived from the trigonometric partitions of the chord, where we set on the **x**-axis the integer numbers **n** with their power **s**, and their height on the **y**-axis with the variable **m**, we can obtain the modulus **r** and its argument φ. By equating these equations of the Pythagorean triangle with those of the trigonometric partitions to analyze the initial radius, and finally analyzing its nth power **s**, we obtain an equation of the initial radius raised to its power **s**. We observe that the graph of the coordinates **x** and **y**, when **n** varies with the angle over time, takes on negative values. The plots of the coordinates **x** and **y** then disappear in the 2D and 3D planes, and the parametric equations become point plots. In other cases, these points in the 2D and 3D planes correspond to another equation of points, arcs, circles, or in the 3D plane, the radiograph of a solid of revolution.

$r_0 = r * e^{-2^{p+2} * \sin^2(\varphi/2^{p+1})}$

$r_0^s = r^s * e^{-2^{p+2} * s * (\sin^2(\varphi/2^{p+1}))}$

$r = \sqrt{(n^{2s} + m^2)}$

$r^s = \sqrt{(n^{2s} + m^2)^s}$, substituting **r** and renaming r_0^s;

$R_0 = \sqrt{(n^{2s} + m^2)^s} * e^{-2^{p+2} * s * (\sin^2(\varphi/2^{p+1}))}$, whose components **x** and **y** are:

$R_{0x} = \cos(k * \varphi) * \sqrt{(n^{2s} + m^2)^s} * e^{-2^{p+2} * s * (\sin^2(\varphi/2^{p+1}))}$

$R_{0y} = \sin(k * \varphi) * \sqrt{(n^{2s} + m^2)^s} * e^{-2^{p+2} * s * (\sin^2(\varphi/2^{p+1}))}$

Figure 2.52: Graph showing the relationship between irrational and complex numbers.

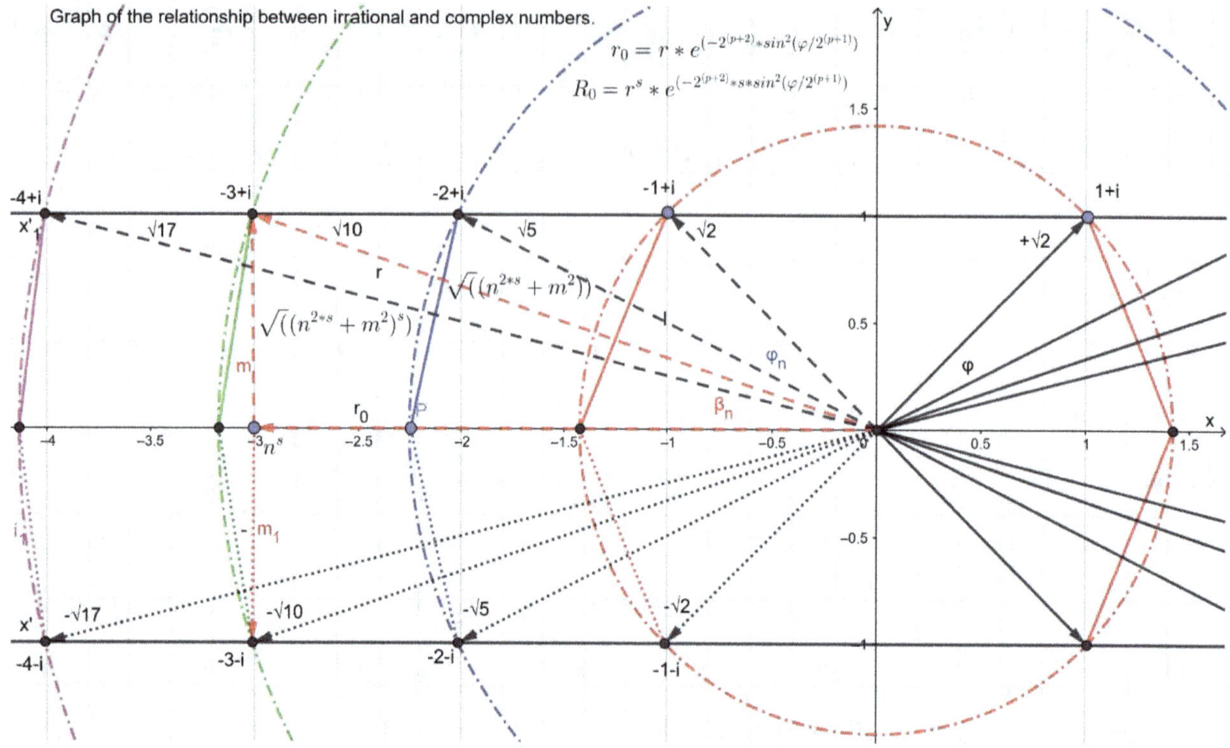

Note: This diagram allows the analysis of the relationship between irrational and complex numbers using the various equations of trigonometric partitions.

Figure 2.53: Graph of the initial radius of the trigonometric partitions of the chord.

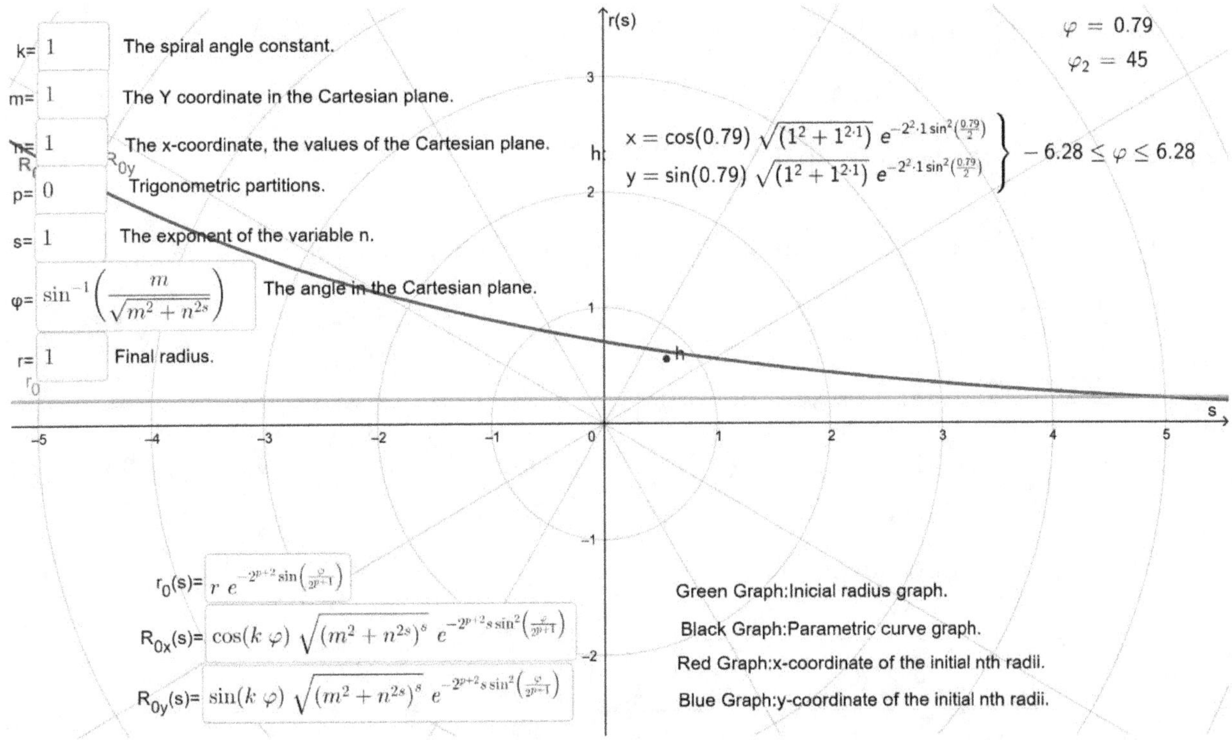

Note: Relationship between the modulus, argument, and the equations of the trigonometric partitions of the chord.

In these equations, the choice of the angle equation used within the trigonometric partition formulas is of utmost importance, as it determines the resulting figure generated by the formula. Different angle equations yield distinct outcomes when applied to the mathematical equations of the trigonometric partitions.

Figure 2.54: Polar curve of the initial radius of the trigonometric partitions of the chord.

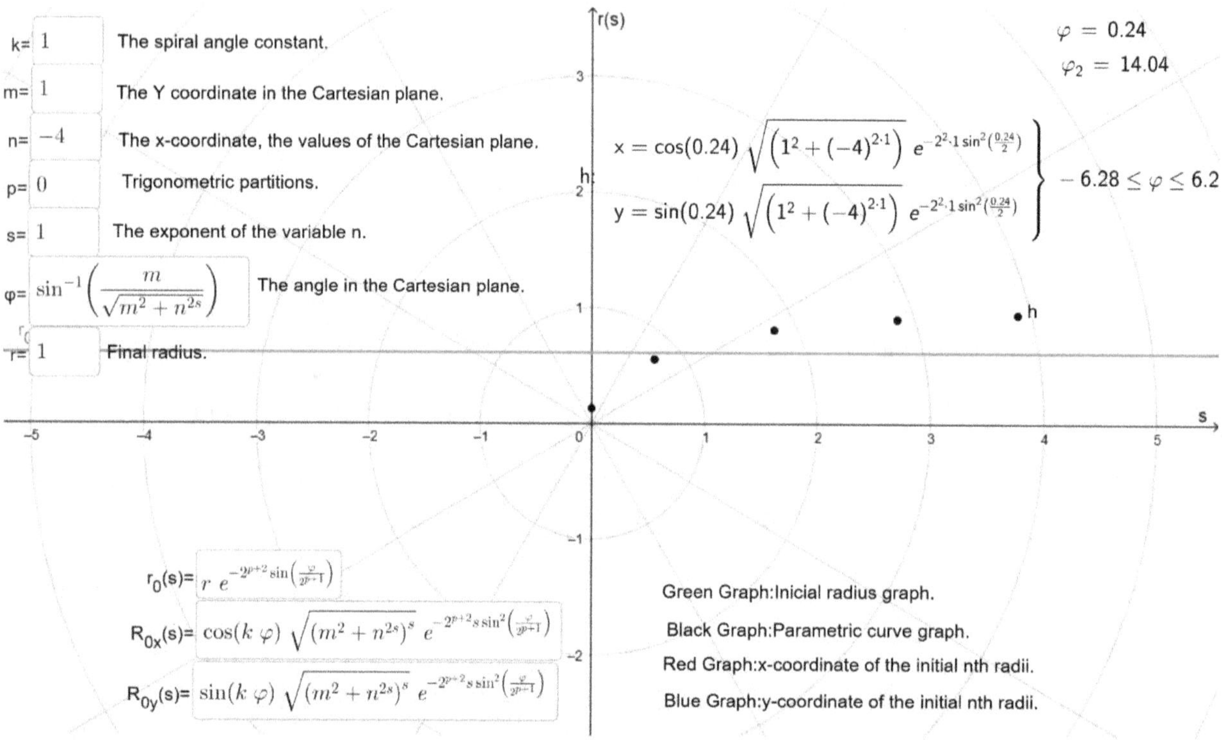

Note: Polar curve illustrating the relationship between the modulus, argument, and the equations of the trigonometric partitions of the chord.

This same technique can be used to construct the Theodorus or Pythagorean spiral, which can be generated by nodes. There are two ways to create the Pythagorean spiral. The first uses an equation developed by the Theory of Spiral Angles, Spirals, and Trigonometric Partitions, presented in the first book on page 36. The second uses a general equation to analyze all radical spirals of nth index, producing an egg-shaped figure in 3D space. When the root index tends to infinity, the resulting figure is a circle in 2D and a sphere in 3D. The polar equation using the angles between radii, called spiral angles, generates the straight line of points or nodes (blue line). The polar equation with the summation of spiral angles between the radii generates the spiral of nodes (red line). In GeoGebra, the domain of the polar function, controlled by the sliders for the values of **n** and **f**, must automatically take the same values to generate the red spiral by nodes, since **f** represents the increment added to each spiral angle generated between radii. These equations apply specifically to the square root.

$\varphi = 2 * \sin^{-1}\sqrt{(\Delta r/d)}$. Spiral angle.For the blue graph.

$\beta = \sum_{n=1}^{f} \varphi_n$. Summation of angles between radii, related to the number of nodes.

β is for the red graph.

$d = 2 * r$. Diameter.

$c = \sqrt{(\Delta r * d)}$. Chord.

$s = (\pi/180) * d * \sin^{-1}\sqrt{(\Delta r/d)}$. Circular sector.

$\varphi_n = 2 * \sin^{-1}\sqrt{((\sqrt{(n+1)} - \sqrt{n})/(2 * \sqrt{(n+1)}))}$

$s_n = 2 * \sqrt{(n+1)} * \sin^{-1}\sqrt{((\sqrt{(n+1)} - \sqrt{n})/(2 * \sqrt{(n+1)}))}$

$c_n = \sqrt{2 * (\sqrt{(n+1)}) * (\sqrt{(n+1)} - \sqrt{n})}$

Figure 2.55: Graph of the Theodorus spiral by nodes, based on the square root.

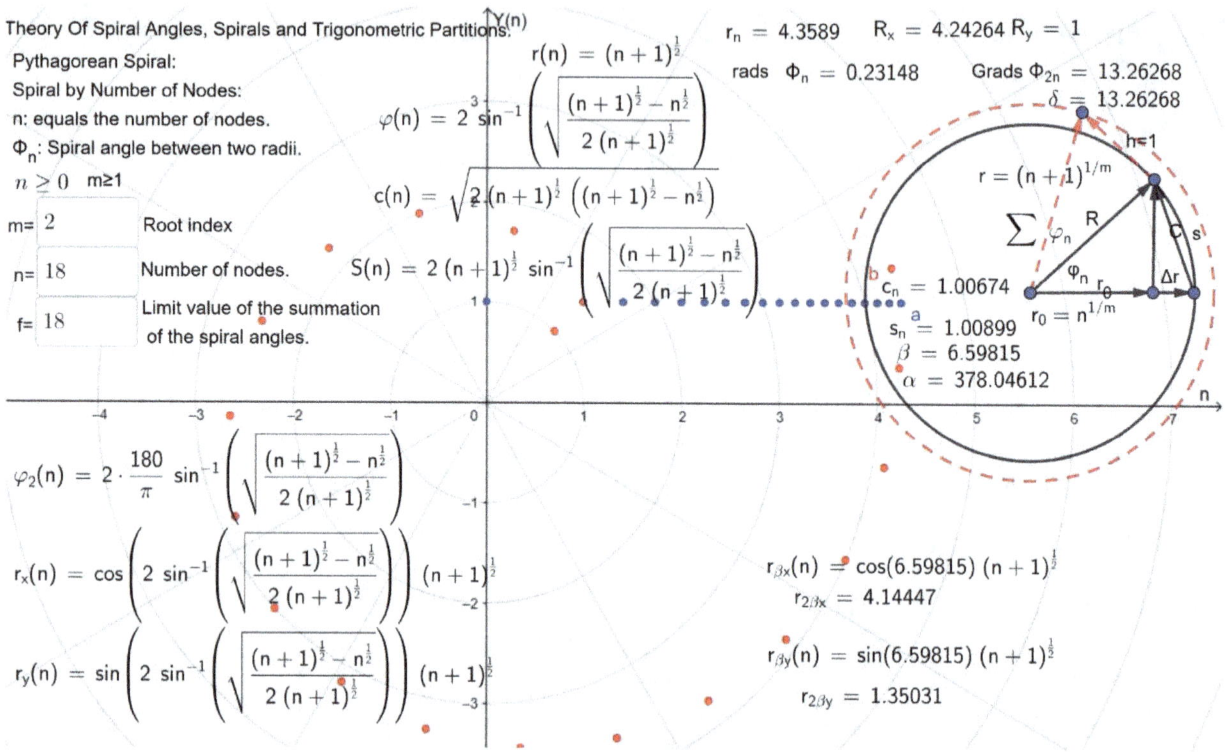

Note: Polar curves **a** and **b** in blue and red generate the blue straight line and the red spiral by nodes.

These equations can be used to analyze spirals of irrational numbers with all radical numbers, with the radicand index extending to infinity.

The radii can be substituted into all equations of the circle's components analyzed in spiral form, with their corresponding spiral angles. The equations independent of the radius are:

$r = \csc(\varphi)$. Radius in trigonometric spiral form, with its spiral angle.

$\beta = \sum_{n=1}^{f} \varphi_n$. Summation of the angles between radii, related to the number of nodes.

φ is for the blue graph.

β is for the red graph.

$r = \sqrt{(n+1)}$. Radius by nodes.

$\Delta s = \varphi * \csc(\varphi)$. Circular sector.

$c = \sec(\varphi/2)$. Chord.

$a = 0.5 * \csc(\varphi/2)$. Apothem.

$f = 0.5 * \sec(\varphi/2) * \tan(\varphi/4)$. Arrow.

$\Delta r = \tan(\varphi/2)$. Radius growth.

Figure 2.56: Graph of the cubic spiral by nodes.

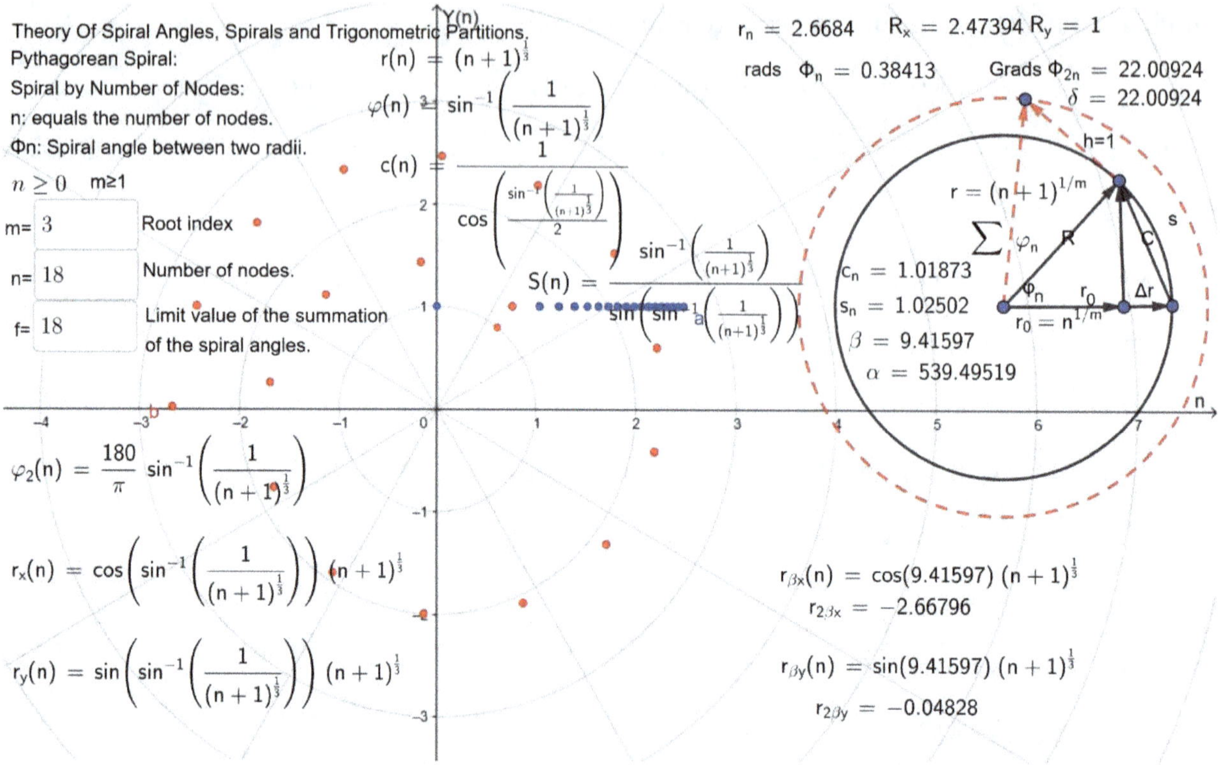

Note: Polar curves **a** and **b** in blue and red generate the blue straight line and the red node spiral for the cubic root. These equations can be used to analyze all root-based spirals.

Figure 2.57: Graph of the cubic spiral in the 3D plane, rotating along the **x**-axis.

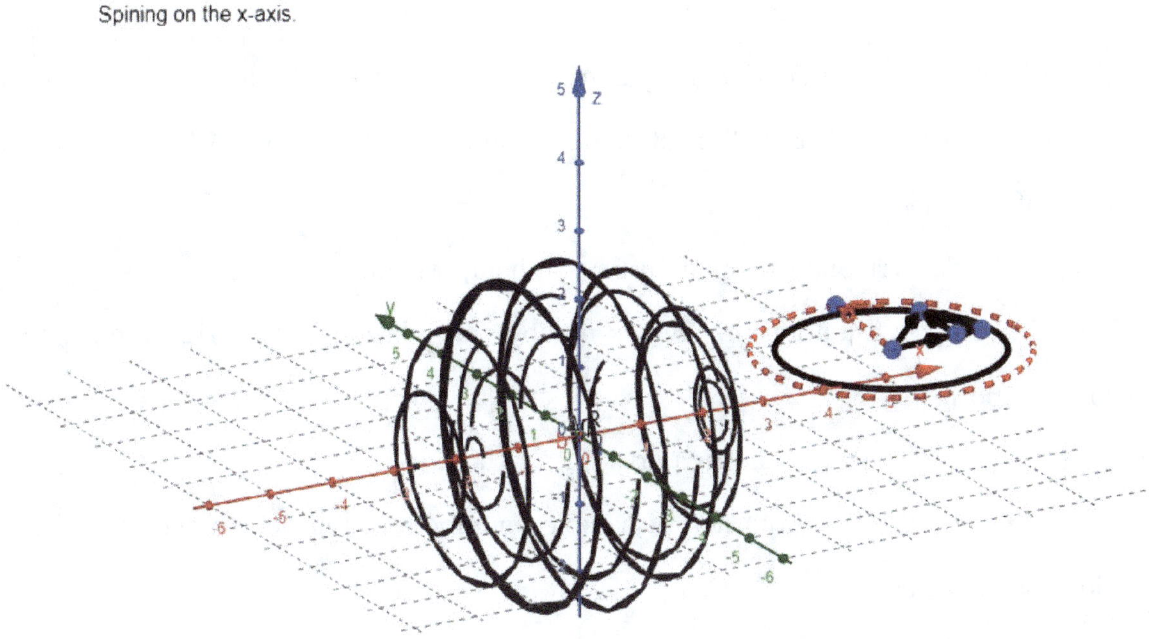

Note: Polar curves **a** and **b** in blue and red generate the blue straight line and the red node spiral in the 2D plane, and in the 3D plane, the red spiral generates an egg-shaped form.

The mathematical equation of Leonhard Euler for complex numbers.

$r = r_0 * e^{i*\varphi} = r_0 * (\cos\varphi + i * \sin\varphi)$, or $r = r_0 * e^{-i*\varphi} = r_0 * (\cos\varphi - i * \sin\varphi)$.

This concept also applies to De Moivre's Theorem:

$(\cos\varphi + i * \sin\varphi)^n = \cos(n * \varphi) + i * \sin(n * \varphi)$.

This equation can be related to all the equations of the trigonometric partitions, allowing us to find the equivalences of complex numbers in relation to each of them. This helps us understand the different values that complex numbers can take on the unit circle. In these equations, all equivalent expressions of the angle can be used, for example, $\varphi = \omega * t = 2 * \pi * f * t = k * x - \omega * t$, with $k = 2 * \pi/\lambda$..It can even be applied to small angles, using the equation $\varphi \approx 2 * \sin(\varphi/2)$.

If we equate the radii of the equations to solve for the complex numbers, we obtain the following results.

$r = r_0 * e^{i*\varphi}$. Equation of complex numbers from the wave function.

$r = r_0 * e^{2^{p+2}*\sin^2(\varphi/2^{p+1})}$. Trigonometric partitions of the chord.

$i = 2^{p+2} * \sin^2(\varphi/2^{p+1})/\varphi$. Equation of complex numbers in relation to the trigonometric partitions of the chord.

$r = r_0 * e^{2^p*\sin(\varphi/2^p)}$. Trigonometric partitions of the apothem.

$i = 2^p * \sin(\varphi/2^p)/\varphi$. Equation of complex numbers in relation to the trigonometric partitions of the apothem.

$r = r_0 * e^{\varphi-2^p*\sin(\varphi/2^p)}$. Trigonometric partitions of the arrow.

$i = [\varphi - 2^p * \sin(\varphi/2^p)]/\varphi$. Equation of complex numbers in relation to the trigonometric partitions of the arrow.

$r = r_0 * e^{\varphi-2^{p-1}*\sin(\varphi/2^{p-1})}$. Trigonometric partitions of the radial growth.

$i = [\varphi - 2^{p-1} * \sin(\varphi/2^{p-1})]/\varphi$. Equation of complex numbers in relation to the trigonometric partitions of the radial growth. And so on.

If we convert these equations of complex numbers i as a function of the angle, and introduce them into the radius equation, we can continue generating other types of wavelets.

$r = r_0\,(\cos\varphi + i(\varphi)\,\sin\varphi), \quad r = r_0\,e^{i(\varphi)\varphi}$

$i(\varphi) = \dfrac{2^{p+2}\,\sin^2(\varphi/2^{p+1})}{\varphi}.$

$r = r_0\left(\cos\varphi + \left[\dfrac{2^{p+2}\,\sin^2(\varphi/2^{p+1})}{\varphi}\right]\sin\varphi\right)$.Complex numbers can be converted into a wave function.

If the radio compression is used: $i(\varphi) = \dfrac{-2^{p+2}\,\sin^2(\varphi/2^{p+1})}{\varphi}.$

$r = r_0\left(\cos\varphi - \left[\dfrac{2^{p+2}\,\sin^2(\varphi/2^{p+1})}{\varphi}\right]\sin\varphi\right).$

$r_x = r_0 * \cos(k*\varphi)\left(\cos\varphi - \left[\dfrac{2^{p+2}\,\sin^2(\varphi/2^{p+1})}{\varphi}\right]\sin\varphi\right).$

$r_y = r_0 * \sin(k*\varphi)\left(\cos\varphi - \left[\dfrac{2^{p+2}\,\sin^2(\varphi/2^{p+1})}{\varphi}\right]\sin\varphi\right).$

"Likewise, wavelet transformations can be performed using the equivalent angle formulas, allowing the generation of other types of wavelets through all the equations of the trigonometric partitions." Imaginary or complex numbers also represent a function, and within a mathematical equation, they can produce different behaviors, resulting in a wide variety of graphs or solutions. For example:In Riemann Z function $z = n(\Delta L)^{1/s}$.

Where ΔL is the difference between the largest prime number and the smallest one.

$F(z) = e^z = F(n(\Delta L)^{1/s}) = e^{n(\Delta L)^{1/s}}$ with $s = 1$, $F(z) = z^2 = F(n(\Delta L)^{1/s}) = n(\Delta L)^{1/s}$ with $s = \frac{1}{2}$, $F(z) = z^3 = F(n(\Delta L)^{1/s}) = n(\Delta L)^{1/s}$ with $s = \frac{1}{3}$,

$F(z) = 1/z = F(n(\Delta L)^{1/s}) = 1/n(\Delta L)^{1/s}$,with $s = 1$ and so on.

It can be observed that when solving the equations in relation to the trigonometric partitions and Euler's equation, the trigonometric SINC function appears as a result.

Figure 2.58: Graph of the Equations of the trigonometric partitions of the chord with the radius in compression,depending on the angle.

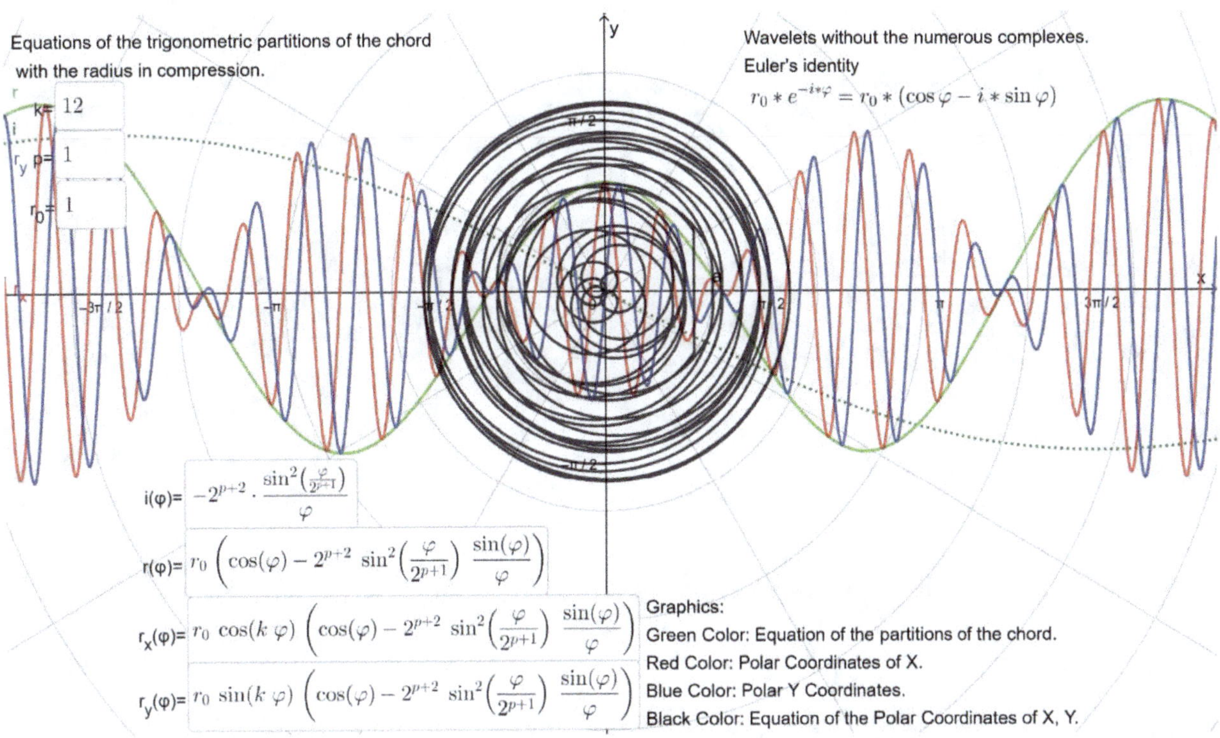

Note: Graphs of the wavelets using the mathematical equations of complex numbers in relation to the trigonometric partition equations of the chord.

Figure 2.59: Graph of the Equations of the trigonometric partitions of the chord with the radius in compression, depending on the frequency and time.

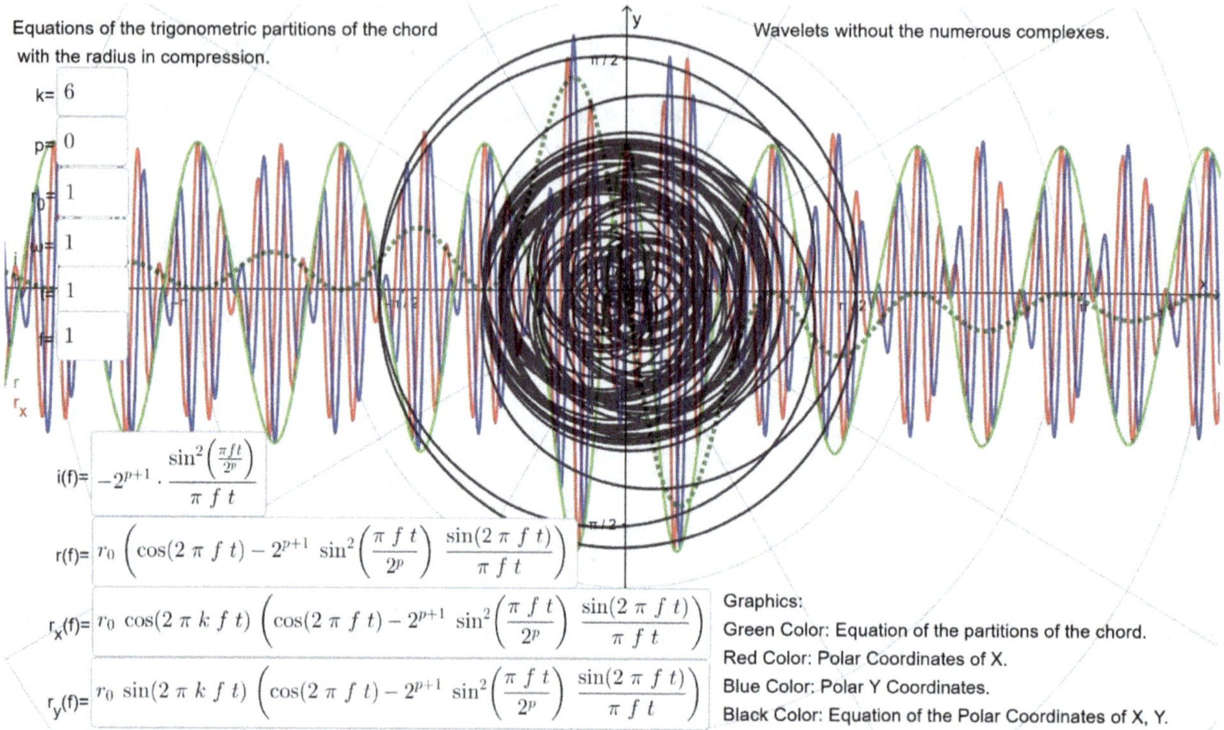

Note: Graphs of the wavelet transforms using the mathematical equations of complex numbers in relation to the trigonometric partition equations of the chord.

Figure 2.60: Graph of the Equations of the trigonometric partitions of the chord with the radius in compression,depending on the position, time, angular velocity, and time.

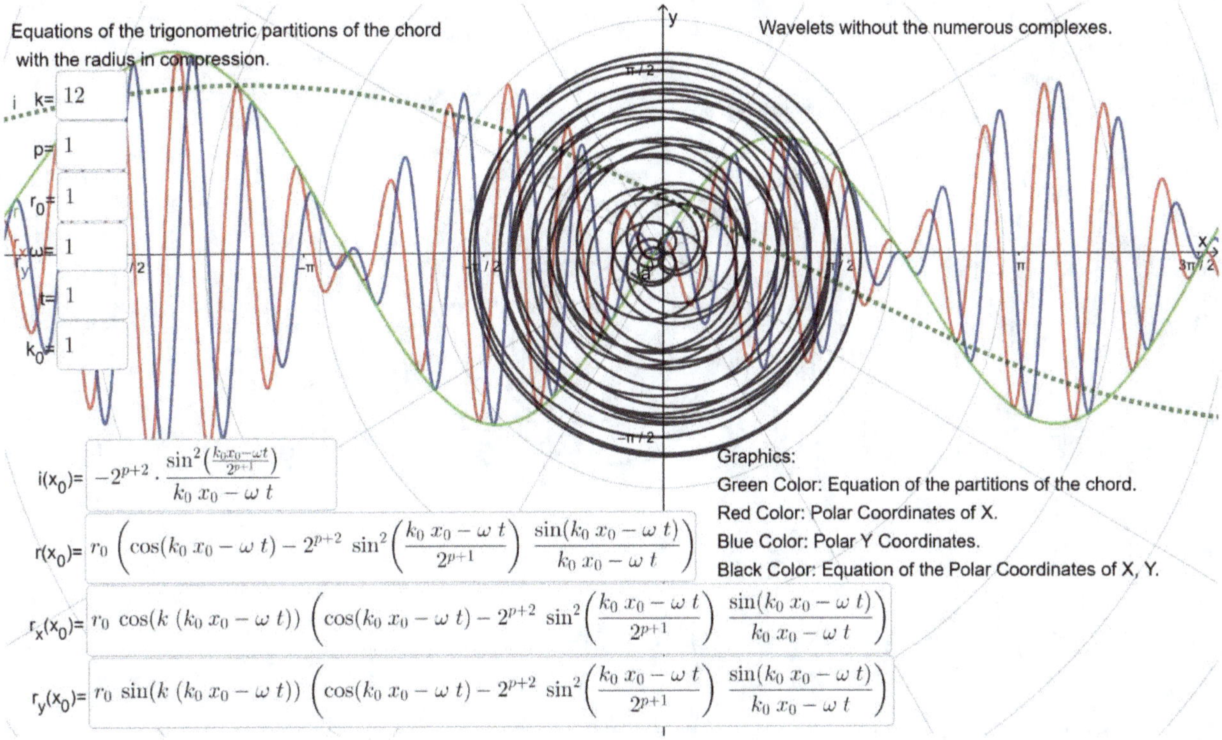

Note: Graphs of the wavelet transforms using the mathematical equations of complex numbers in relation to the trigonometric partition equations of the chord.

Figure 2.61: Graph of the Equations of the trigonometric partitions of the chord with the radius in compression,depending on the wavelength,position, time, angular velocity, and time.

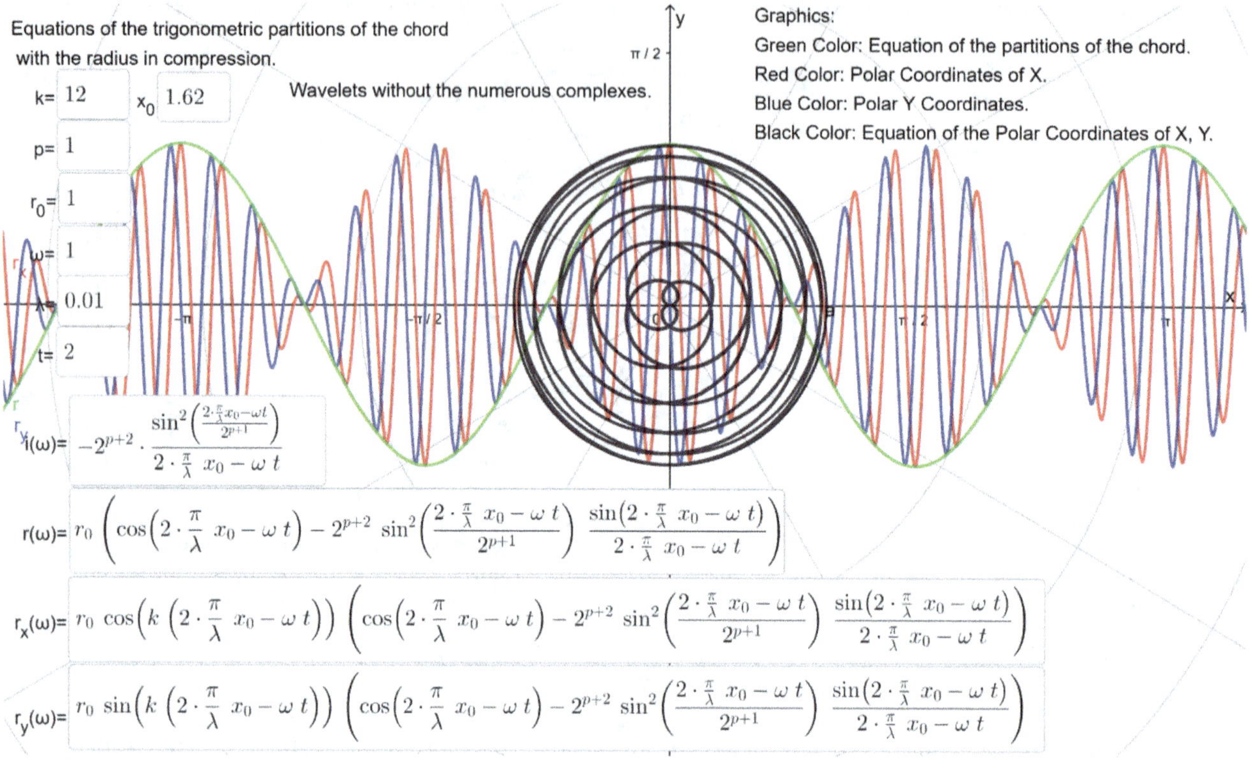

Note: Graphs of the wavelet transforms using the mathematical equations of complex numbers in relation to the trigonometric partition equations of the chord.

When a mathematical equation that generates a wavelet is subjected to a partial derivative, the resulting derivative equation produces another wavelet of greater magnitude, and this process can continue successively. Using the equations of complex numbers in relation to the trigonometric partitions of the components of a circle, it is possible to generate wavelets from the partial derivatives of the complex-number equations equivalent to the trigonometric partitions. These same concepts can also be applied to Erwin Schrödinger's equation, analyzing it through the equivalent complex-number equations and the partial derivatives of the trigonometric partitions.

$i * \frac{\partial y}{\partial t} = \frac{\hbar}{2*m} * \frac{\partial^2 y}{\partial x^2}$

The wavelet graphs generated by the equations of Euler's identity and De Moivre's theorem are not the same. De Moivre's theorem can likewise be applied to all the transformations of the equivalent angle equations.

Figure 2.62: Graph of the Equations of the trigonometric partitions of the chord with the radius in stretching,depending on the angle.

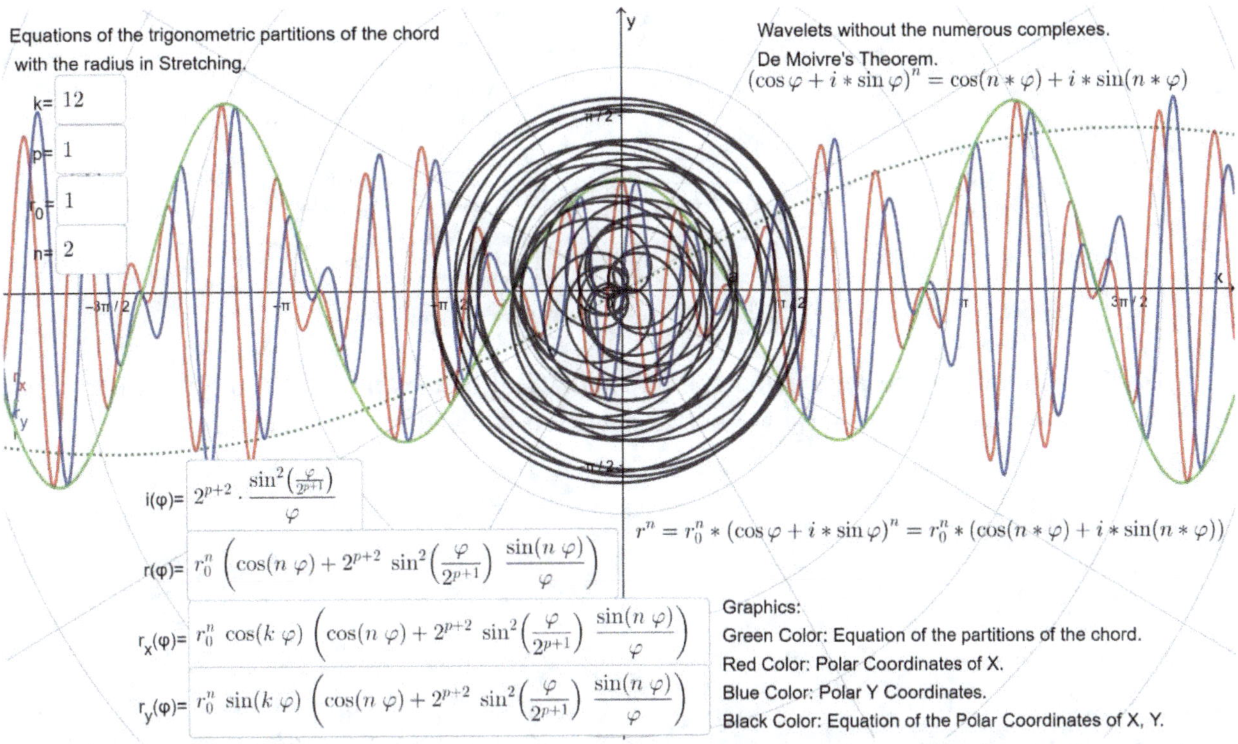

Note: Graphs of the wavelets using the mathematical equations of Moivre's theorem in relation to the trigonometric partition equations of the chord.

We can also generate wavelets using the equivalent angle equation equal to $\varphi = 2*\pi*$n, where **n** represents the number of turns that can be made within a circle.The equations of complex numbers **i** and radii must be expressed as a function of **n**.

Figure 2.63: Graph of the Equations of the trigonometric partitions of the chord with the radius in stretching,Schrödinger's Equation.

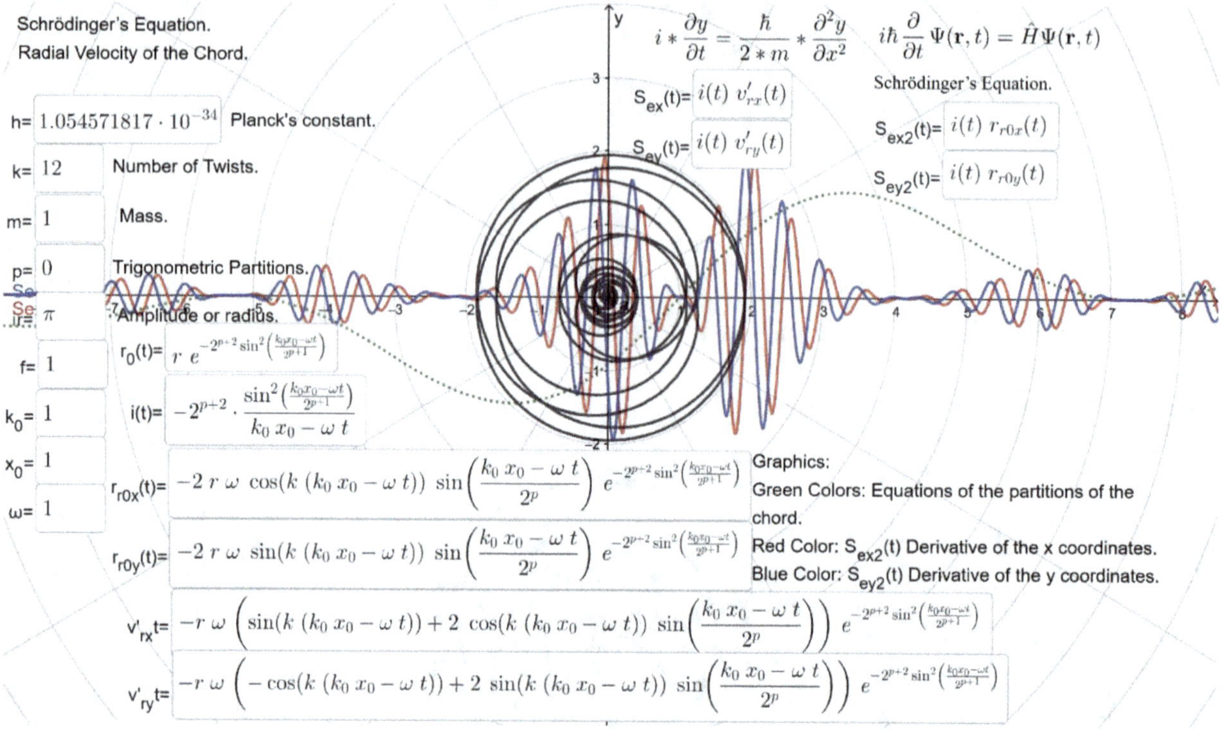

Note: Graphs of the wavelets using the mathematical equations of Erwin Schrödinger in relation to the trigonometric partition equations of the chord.

If we apply this same concept in two ways: First, by taking the partial derivatives of the radius with respect to time and obtaining the x and y components of the derivative of the radius, which represents the radial velocity. Second, by taking the second partial derivatives of the equations of the x and y coordinates of the trigonometric partitions with respect to time. By applying these two processes to the wave equation, wavelets continue to be generated.

$$\frac{\partial^2 y}{\partial t^2} = c^2 \frac{\partial^2 y}{\partial x^2}$$

Figure 2.64: Graph of the Equations of the trigonometric partitions of the chord with the radius in compression,derivative of the equation of complex numbers as a function of time and frecuency.

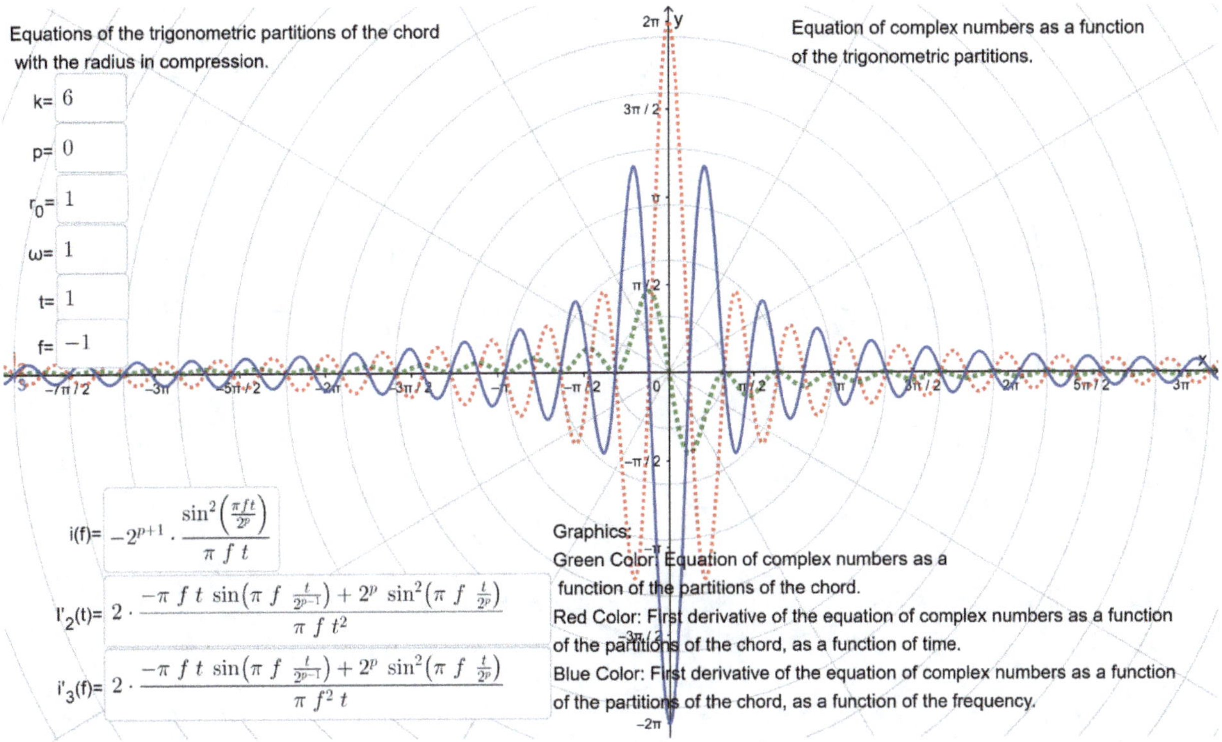

Note: Graphs of the wavelets using the mathematical equations of complex numbers in relation to the trigonometric partition equations of the chord.

Figure 2.65.Two ways to derive the equations trigonometric partitions.

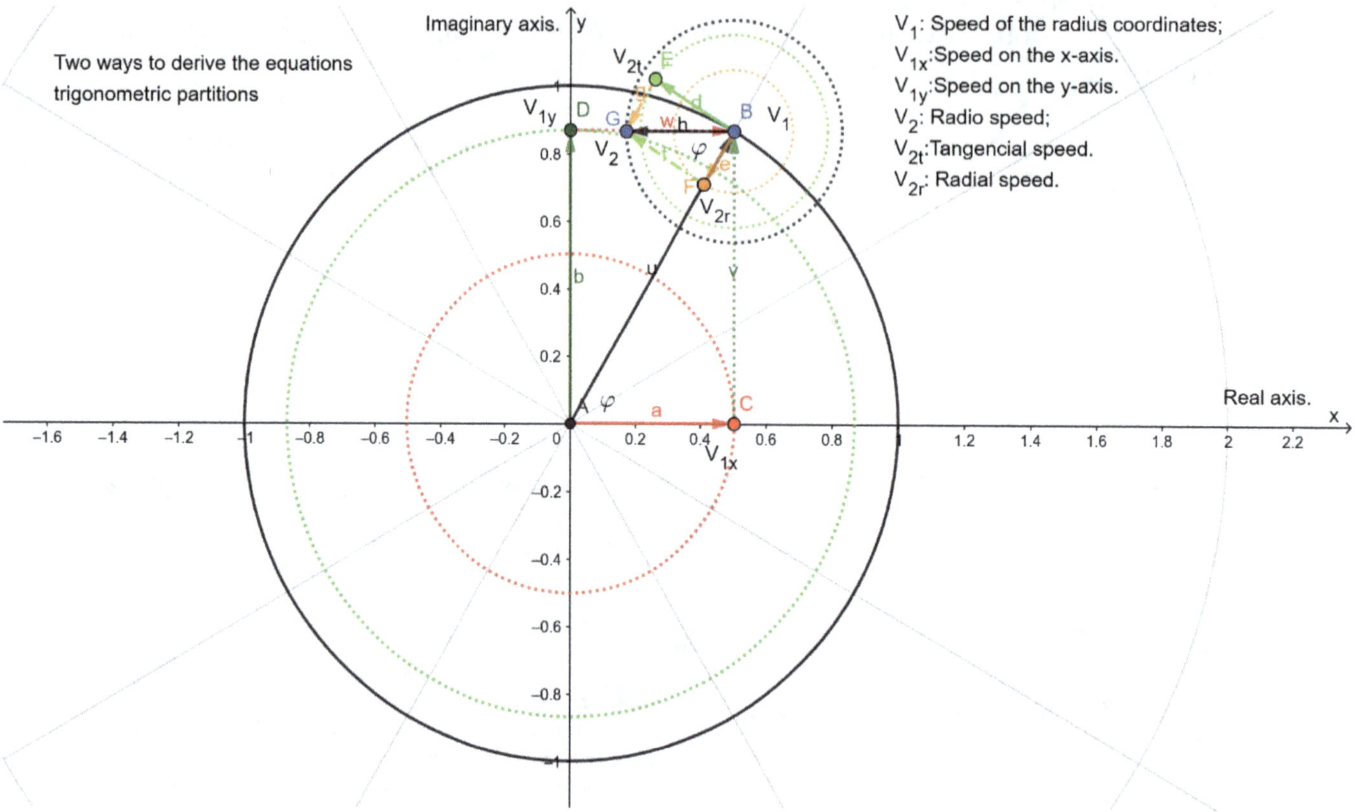

Note: These are the two methods for deriving the equations of trigonometric partitions to analyze equations that contain both real and complex numbers.

2.14 Annexes: List Of Videos Of The Wavelets With The Theory Of Trigonometric Partition:

https://youtube.com/shorts/gNe9x_lmqCI?feature=share

https://youtube.com/shorts/tBzZuS9apIk?feature=share

https://youtube.com/shorts/nPy5xm44dqY?feature=share

https://youtube.com/shorts/Kc5NPw6YUvg?feature=share

https://youtu.be/BCTKeLcrhtg

https://youtu.be/hp4Hyx9CN0w

https://youtu.be/9dqpQ1QCGtY

https://youtu.be/Od0dJIyF9rk

https://youtu.be/v1NINH9tzyk

https://youtu.be/_0GXbzWL6LI

https://youtu.be/N9DQJKtGfiU

https://youtu.be/GPQLfEsh0t0

https://youtu.be/GilCFkLWiJE

https://youtu.be/6LpoYQS1a7g

https://youtu.be/BQOKpLiaGH0

https://youtu.be/3Ebvypj577E

https://youtu.be/DV1iJV0oa7Y

https://youtu.be/p0Zc9onKQ0Q

https://youtu.be/TvqSpJv7RqE

https://youtu.be/pH-OCcdHP84

https://youtu.be/ZsHNJjjYspk

https://youtu.be/zTQHj0EhNmM

https://youtu.be/8qzi5Qbx-So

https://youtu.be/4HyT8fd6zTM?si=fJOSLj8yz2BZgqcV

2.15 Social Networks

YouTube:

https://www.youtube.com/@josephdays07

Linked In:

https://www.linkedin.com/in/jos%C3%A9-mauricio-orellana-d%C3%ADaz-71b671200

Instagram:

https://www.instagram.com/josemauricio.orellanadiaz?r=nametag

TikTok:

tiktok.com/@josephdays07

References

Fernández, A. O. (2019). Ondículas, evolución de algunas ideas y aplicaciones. *Selecciones Matemáticas*, *6*(1), 119–127.

José Mauricio Orellana Díaz. (n.d.-a). *Ondículas de potencias enésimas.wavelets of nth powers.* https://youtu.be/zTQHj0EhNmM?si=xZJhjntalYaAdzRw.

José Mauricio Orellana Díaz. (n.d.-b). *Ondículas en función de la frecuencia, la velocidad angular y el tiempo. wavelet transforms.* https://youtu.be/ZsHNJjjYspk?si=iO$_T$*tSSeGLqSHGyF.*

José Mauricio Orellana Díaz. (n.d.-c). *Primera derivada de las particiones de la cuerda, y el gráfico del principio de incertidumbre.* https://youtu.be/pH-OCcdHP84?si=gyHG-4kqkbieyjd5.

Orellana Díaz, J. M. (2023). *Theory of spiral angles, spirals and trigonometric partitions* (Vol. 1). Barker and Joules.

Serway, R. A., et al. (2018). Fundamentos de física.

www.ingramcontent.com/pod-product-compliance
Lightning Source LLC
LaVergne TN
LVHW060059220926
848484LV00008BA/2374